Drug Interactions Casebook

The Cytochrome P450 System and Beyond

Drug Interactions Casebook

The Cytochrome P450 System and Beyond

Neil B. Sandson, M.D.
Director, Division of Education and Residency Training,
Sheppard Pratt Health System, Towson, Maryland;
Associate Director, University of Maryland/
Sheppard Pratt Psychiatry Residency Program;
Clinical Assistant Professor, Department of Psychiatry,
University of Maryland Medical System,
Baltimore, Maryland

with contributions by

Kelly L. Cozza, M.D.
Scott C. Armstrong, M.D.
Jessica R. Oesterheld, M.D.

Washington, DC
London, England

Note: The authors have worked to ensure that all information in this book is accurate at the time of publication and consistent with general psychiatric and medical standards, and that information concerning drug dosages, schedules, and routes of administration is accurate at the time of publication and consistent with standards set by the U. S. Food and Drug Administration and the general medical community. As medical research and practice continue to advance, however, therapeutic standards may change. Moreover, specific situations may require a specific therapeutic response not included in this book. For these reasons and because human and mechanical errors sometimes occur, we recommend that readers follow the advice of physicians directly involved in their care or the care of a member of their family.

Books published by American Psychiatric Publishing, Inc., represent the views and opinions of the individual authors and do not necessarily represent the policies and opinions of APPI or the American Psychiatric Association.

Manufactured in the United States of America on acid-free paper
07 06 05 04 03 5 4 3 2 1
First Edition

Typeset in Adobe's New Baskerville and Helvetica

American Psychiatric Publishing, Inc.
1000 Wilson Boulevard
Arlington, VA 22209-3901
www.appi.org

Library of Congress Cataloging-in-Publication Data
Sandson, Neil B., 1965–
Drug interactions casebook : the cytochrome P450 system and beyond / Neil B. Sandson.—1st ed.
p. ; cm.
Includes bibliographical references and index.
ISBN 1-58562-091-2 (alk. paper)
1. Psychotropic drugs—Side effects—Case studies. 2. Drug interactions—Case studies. 3. Cytochrome P-450. I. Title: Cytochrome P450 system and beyond. II. Title.
[DNLM: 1. Cytochrome P-450 Enzyme System—metabolism—Case Report. 2. Drug Interactions—Case Report. QV 38 S221d 2003]
RM315.S234 2003
615'.788—dc21

2003043666

British Library Cataloguing in Publication Data
A CIP record is available from the British Library.

Contents

Chapter 3

Chapter 7

Chapter 8

Contributors

Scott C. Armstrong, M.D.

Medical Co-Director, Tuality Center for Geriatric Psychiatry, Forest Grove, Oregon; Associate Clinical Professor of Psychiatry, Oregon Health and Science University, Portland, Oregon

Kelly L. Cozza, M.D.

Psychiatrist, Infectious Disease Service, Department of Medicine, Walter Reed Army Medical Center, Washington, D.C.; Assistant Professor, Department of Psychiatry, Uniformed Services University of the Health Sciences, Bethesda, Maryland; Fellow, Academy of Psychosomatic Medicine

Jessica R. Oesterheld, M.D.

Medical Director, Spurwink School, Portland, Maine; Instructor in Family Medicine, University of New England School of Osteopathy, Biddle, Maine

Neil B. Sandson, M.D.

Director, Division of Education and Residency Training, Sheppard Pratt Health System, Towson, Maryland; Associate Director, University of Maryland/Sheppard Pratt Psychiatry Residency Program; Clinical Assistant Professor, Department of Psychiatry, University of Maryland Medical System, Baltimore, Maryland

Introduction

I've been teaching psychiatry residents for the past 7 years. When this teaching has been directed toward clinical matters, these residents have engaged the material quite readily. A depressed or psychotic patient can serve as an organizing focus for specific teaching points. However, when the subject has turned to drug interactions, especially the much-feared cytochrome P450 system, I've encountered anxiety reactions that run the gamut from avoidance to dissociation. Eyes glaze over, drug company pens are chewed on, and little is retained at the end of the day. I have long sought an effective way to present this critically important material so that it will feel more accessible and interesting.

That has been the organizing theme of this book. My hope is that a case-series format will bring the material to life so that readers can retain pearls and apply them to their patients. I also hope that this book will appeal to not only psychiatrists but also a wide range of medical specialists and generalists, as well as those who are in any way acquainted with the mental health field. Almost every case contains at least one psychoactive agent, but there is plenty of material from other branches of medicine to be found herein.

Every case in this book is grounded in reality. None of these interactions are merely the product of theoretical musings or the pairing of substrates and inhibitors of the same P450 enzyme according to some table or chart. Many of the cases are direct narrations of events that have happened to patients of mine or of my colleagues, with the identifiers subtly altered to ensure confidentiality. Others are case constructions around well-known, accepted, and multiply verified interactions. Regardless of the precise nature of the source(s), all cases in this book can be relied on for their essential accuracy in their depictions of real clinical events. Some of the interactions depicted may occur only infrequently when the medications in question are combined, but this will be mentioned in the discussion section following each case. Every individual is different, and the differences in individual metabolic capabilities are vast, with some estimates citing 10- to 30-fold variations in 3A4 activity that are nonpathologic (Ketter et al. 1995). Hence, only a subset of these interactions will necessarily occur in most or all individuals who encounter these medication combinations. However, unless otherwise specified, there is a high enough likelihood of depicted interactions occurring that cautious clinicians would be well advised to prospectively anticipate and plan for them.

This book is not designed to comprehensively address the topic of the cytochrome P450 system as an emerging field of study. That has been ably accomplished by Drs. Kelly Cozza, Scott Armstrong, and Jessica Oesterheld in their *Concise Guide to Drug Interaction Principles for Medical Practice: Cytochrome P450s, UGTs, P-Glycoproteins,* 2nd Edition (Cozza et al. 2003). This book has actually been conceived and written as an adjunct to that work, complementing the breadth of information contained therein with a wide array of cases that illustrate the principles detailed in that book.

As engaging as a case series may be, I have tried to make this book useful to the reader who does not wish to peruse it cover to cover over a slow weekend. There is a thorough index, which will enable readers to find information relating to specific medications by both generic and trade names. The index will also classify each of the cases according to which of the six basic P450 interaction patterns it illustrates (see Chapter 1: "Core Concepts"). The chapters are organized by the nature of the mechanistic processes that account for the depicted drug interactions ("2D6 Case Vignettes," "3A4 Case Vignettes," "Non-P450 Case Vignettes," etc.). Also, Drs. Cozza, Armstrong, and Oesterheld have generously agreed to allow the P450 tables from their concise guide to be included among the appendixes of this book. This will allow interested readers to extrapolate to specific interactions that are possible, likely, or even virtually certain but that have not been specifically depicted in this book.

For better or worse, the world of drug interactions is becoming more complicated as our understanding of drug metabolism expands. In this spirit, the first chapter ("Core Concepts") also contains brief explanations about phase II glucuronidation and P-glycoproteins, and the appendixes contain tables (generously donated by Dr. Oesterheld) that allow the reader to derive, anticipate, and explain possible interactions that involve these systems.

I welcome you to this exploration of the world of drug interactions. For those intrepid souls among you, I have included a self-assessment test case at the end of the book so that you can evaluate your progress in becoming a drug interaction detective. Good luck and happy hunting!

References

Cozza KL, Armstrong SC, Oesterheld JR: Concise Guide to Drug Interaction Principles for Medical Practice: Cytochrome P450s, UGTs, P-Glycoproteins, 2nd Edition. Washington, DC, American Psychiatric Publishing, 2003

Ketter TA, Flockhart DA, Post RM, et al: The emerging role of cytochrome P450 3A in psychopharmacology. J Clin Psychopharmacol 15(6):387–398, 1995

Acknowledgments

This book is dedicated to my two wonderful children, Charlie and Lisa. Heartfelt thanks to my patient wife and all my friends, family, and coworkers who provided moral and material support. This was a team effort, and you made it possible. Special thanks also to Dr. Steven Sharfstein for his patronage and guidance; Dr. Sunil Khushalani for his invaluable assistance; Drs. Cozza, Armstrong, and Oesterheld for their warm welcome into the "P450 family" and ongoing mentorship; and my parents for pointing me in the right directions.

I would also like to thank the following people who contributed clinical vignettes:

- Scott Armstrong, M.D.
- Daniela Boerescu, M.D.
- Kelly Cozza, M.D.
- Gabriel Eckermann, M.D.
- Sunil Khushalani, M.D.
- Christian Lachner, M.D.
- Michael Levinson, M.D.
- Lily Lin, Pharm.D.
- Raymond Love, Pharm.D.
- Catherine Marcucci, M.D.
- Jessica Oesterheld, M.D.
- Steven Ruths, M.D.
- Joseph Sokal, M.D.
- Susan Strahan, M.D.
- Ken Walters, Pharm.D.

Chapter 1

Core Concepts

In this chapter, I provide concise explanations of the concepts and terms that are used throughout the various cases in subsequent chapters. These explanations are designed to enable the P450 neophyte to dive right in and begin reading the cases. This chapter is not intended to provide a comprehensive review of these topics. The interested reader should consult *Concise Guide to Drug Interaction Principles for Medical Practice: Cytochrome P450s, UGTs, P-Glycoproteins,* 2nd Edition (Cozza et al. 2003), which provides such a review.

Core Concepts

P450 system A set of enzymes (usually hepatic, but not always) that metabolize a wide array of endogenous and exogenous substrates in order to detoxify them and then eliminate them from the body. These enzymes perform phase I, or oxidative, metabolism, as opposed to phase II conjugation. Each P450 enzyme is named by a number-letter-number sequence (such as "1A2" or "3A4"). This sequence is the rough equivalent of "family, genus, species" as a way of identifying members of the animal kingdom, and serves to identify specific P450 enzymes. 2C9 and 2C19 are more closely related and share more common substrates than do 2C9 and 3A4. There are more than 40 P450 enzymes, but the several that are examined in this book account for more than 90% of P450 metabolism (Guengerich 1997).

Substrate The agent that is metabolized by an enzyme into a metabolic end product. Usually this results in eventual deactivation of the agent in preparation for elimination from the body. In the rare case of "pro-drugs," these agents are initially inactive and rely on enzymes to be metabolized into active compounds.

Inhibitor An agent that interferes with, or inhibits, the functioning of the enzyme or enzymes that metabolize substrates. Enzymatic inhibition comes in two main varieties. The first, *competitive inhibition,* is when two substrates compete for the same substrate binding site(s) on an enzyme. The competitive inhibitor binds so avidly to this substrate binding site that it effectively displaces the other substrate(s), but the enzyme otherwise functions normally. The second type of inhibition, *noncompetitive, or allosteric, inhibition,* occurs when an agent binds to a nonsubstrate binding site on an enzyme. In so doing, the allosteric inhibitor renders the enzyme less efficient in metabolizing all substrates of that enzyme. The introduction of inhibitors leads to *substrate accumulation* and *decreased metabolite formation.* Inhibitors of P450 enzymes generally act within hours to days, as a function of their half-lives, which determine when they are able to gain access to, and then inhibit, the enzymes in question.

Inducer An agent that causes the liver (or other target organ) to produce more of an enzyme, leading to increased metabolism of the substrates of the induced enzyme. This is a purely quantitative concept. The produced enzyme is no more intrinsically active than otherwise; there is just more of it. In the concept of *enzymatic stimulation,* which is being considered for 3A4, the enzyme is rendered more efficient, aside from its quantity in the liver and/or gut. The introduction of inducers leads to *increased metabolite formation* and *more rapid depletion of substrate(s).* Induction of P450 enzymes can be evident within several days but generally takes 2–3 weeks to reach full effect.

Variability There is vast interindividual variability in enzymatic efficiency (10- to 30-fold differences in 3A4 function, for example) across the world population (Ketter et al. 1995), with some trends within specific ethnic groups. This does not even consider those individuals who lack copies of the genes that code for various P450 enzymes, who are referred to as "poor metabolizers" or "polymorphic" for the specific P450 enzymes in question. By contrast, individuals who have extra copies of P450 genes are referred to as "ultrarapid metabolizers." In general, metabolically normal individuals (called "extensive metabolizers") who are exposed to an agent that strongly inhibits a given P450 enzyme are functionally converted to "poor metabolizers" with regard to that specific P450 enzyme, as long as they are exposed to that agent. Conversely, individuals exposed to an inducer may function as ultrarapid metabolizers for that P450 enzyme. Poor metabolizers/polymorphic individuals will generally have much higher blood levels of a substrate at a given dose than will extensive metabolizers. Although this is counterintuitive, the rate at which poor metabolizers/polymorphic individuals metabolize a substrate of that enzyme is *not* influenced by introducing an inhibitor of that enzyme. There is no way to become an "ultrapoor metabolizer." The inhibitor is redundant.

P-glycoprotein (P-gp) One of a superfamily of transporters that line the gut and the blood-brain barrier. P-glycoproteins play a central role in determining to what degree various substances are absorbed into or excreted from the body. P-glycoprotein is an *extruding* transporter, designed to *remove* substances from enterocytes back into the gut lumen, or from the brain back into the blood. P-glycoprotein has substrates, inhibitors, and inducers. P-glycoprotein substrates are extruded as detailed above. P-glycoprotein inhibitors antagonize this process and lead to retention of P-glycoprotein substrates. P-glycoprotein inducers increase the amount of active P-glycoprotein and thus lead to more extrusion of P-glycoprotein substrates.

UGTs Uridine 5′-diphosphate glucuronosyltransferases, a family of enzymes that perform phase II conjugative metabolism (glucuronidation), which usually follows the phase I oxidative metabolism performed by P450 enzymes or other oxidative metabolic steps. This system functions quite similarly to the P450 system, with several enzymes that also have alpha-numeric labels (1A4, 2B15, etc.), with each enzyme having its own substrates, inhibitors, and inducers.

Patterns of P450 Drug-Drug Interactions

There are six basic patterns of P450 drug-drug interactions.

Pattern 1: An Inhibitor Is Added to a Substrate

This pattern generally results in increases in substrate levels. If the substrate has a low therapeutic index, toxicity may result unless care is exercised (such as closely checking blood levels or lowering substrate doses in anticipation of the interaction).

Example: Paroxetine Is Added to Nortriptyline

Nortriptyline is a 2D6 substrate (Sawada and Ohtani 2001), and paroxetine is a 2D6 inhibitor (von Moltke et al. 1995). The addition of paroxetine impairs the ability of 2D6 to metabolize nortriptyline, leading to an increase in the blood level of nortriptyline.

Pattern 2: A Substrate Is Added to an Inhibitor

This pattern may cause difficulties if the substrate has a low therapeutic index and is titrated according to preset guidelines that do not take into account the presence of an inhibitor. If the substrate is titrated to specific blood levels or to therapeutic effect, or with an appreciation that an inhibitor is present, then toxicity is less likely to arise.

Example: Nortriptyline Is Added to Paroxetine

Nortriptyline is a 2D6 substrate (Sawada and Ohtani 2001), and paroxetine is a 2D6 inhibitor (von Moltke et al. 1995). Since paroxetine inhibits the ability of 2D6 to metabolize nortriptyline, the added nortriptyline generates a significantly higher nortriptyline blood level than would occur if the paroxetine was not already present. Toxicity may result unless appropriate caution is taken.

Pattern 3: An Inducer Is Added to a Substrate

This pattern generally results in decreases in substrate levels. A decrease in levels of the substrate may result in a loss of efficacy of the substrate, unless blood levels are followed and/or the substrate doses are increased in anticipation of the interaction.

Example: Carbamazepine Is Added to Haloperidol

Carbamazepine induces 3A4, 1A2, and phase II glucuronidation (Hachad et al. 2002; Lucas et al. 1998; Parker et al. 1998, Spina et al. 1996), and haloperidol is metabolized by 3A4, 2D6, 1A2, and phase II glucuronidation (Desai et al. 2001; Kudo and Ishizaki 1999). This increase in the available amounts of 3A4 and 1A2 (and possibly particular UGTs as well) leads to more efficient metabolism of the haloperidol and a resulting decrease in the blood level of haloperidol. This may lead to clinical decompensation (Hesslinger et al. 1999).

Pattern 4: A Substrate Is Added to an Inducer

This pattern may lead to ineffective dosing if preset dosing guidelines are followed that do not take into account the presence of an inducer. If the substrate is titrated to specific blood levels or to clinical effect, or with an appreciation that an inducer is present, then dosing is more likely to be effective.

Example: Carbamazepine Is Added to Haloperidol

Carbamazepine induces 3A4, 1A2, and phase II glucuronidation (Hachad et al. 2002; Lucas et al. 1998; Parker et al. 1998, Spina et al. 1996), and haloperidol is metabolized by 3A4, 2D6, 1A2, and phase II glucuronidation (Desai et al. 2001; Kudo and Ishizaki 1999). Thus, when a given dose of haloperidol is administered, the presence of increased amounts of 3A4 and 1A2 (and possibly particular UGTs as well) leads to more efficient metabolism of the haloperidol, thus generating a significantly lower haloperidol blood level at that haloperidol dose than would occur if carbamazepine was not already present (Hesslinger et al. 1999).

Pattern 5: Reversal of Inhibition

A substrate and an inhibitor have been coadministered and equilibria have been achieved, and then the inhibitor is discontinued. This leads to a resumption of normal enzyme activity and generally results in decreases in levels of substrate and increased metabolite formation. This may result in loss of efficacy of the substrate unless blood levels are followed and/or substrate doses are increased in anticipation of the reversal of inhibition.

Example: Cimetidine Is Discontinued in the Presence of Nortriptyline

Cimetidine, a 2D6 inhibitor (Martinez et al. 1999), and nortriptyline, a 2D6 substrate (Sawada and Ohtani 2001), have been coadministered at appropriate doses, yielding an appropriate nortriptyline blood level and clinical efficacy. The cimetidine is then discontinued, resulting in a cessation of 2D6 inhibition. 2D6 is then available to more efficiently metabolize the nortriptyline, leading to a significant decrease in the nortriptyline blood level, possibly producing a subtherapeutic nortriptyline level and a loss of clinical efficacy.

Pattern 6: Reversal of Induction

A substrate and an inducer have been coadministered and equilibria have been achieved, and then the inducer is discontinued. This gradually (over

1–3 weeks) results in decreased amounts of available enzyme, leading to increased levels of substrate and decreased metabolite formation. This may result in substrate toxicity if the substrate has a low therapeutic index, unless blood levels are followed and/or substrate doses are decreased in anticipation of the reversal of induction.

Example: Smoking Is Discontinued in the Presence of Clozapine

A two-pack-per-day cigarette smoker has been taking a stable dose of clozapine, yielding an appropriate clozapine blood level and clinical efficacy. Clozapine is a 1A2 substrate (Eiermann et al. 1997), and cigarette smoking induces 1A2 (Schrenk et al. 1998; Zevin and Benowitz 1999). Smoking is then discontinued, resulting in a cessation of 1A2 induction. This leads to decreased metabolism of clozapine and a resulting increase in the blood level of clozapine, possibly to a toxic degree.

The Exception

When pro-drugs (hydrocodone, tramadol, cyclophosphamide, etc.) are the substrates in question, the clinical concerns are reversed. The pattern 1 concern is loss of efficacy, not toxicity. The pattern 3 concern is toxicity, not loss of efficacy, and so forth.

References

Cozza KL, Armstrong SC, Oesterheld JR: Concise Guide to Drug Interaction Principles for Medical Practice: Cytochrome P450s, UGTs, P-Glycoproteins, 2nd Edition. Washington, DC, American Psychiatric Publishing, 2003

Desai HD, Seabolt J, Jann MW: Smoking in patients receiving psychotropic medications: a pharmacokinetic perspective. CNS Drugs 15(6):469–494, 2001

Eiermann B, Engel G, Johansson I, et al: The involvement of CYP1A2 and CYP3A4 in the metabolism of clozapine. Br J Clin Pharmacol 44(5):439–446, 1997

Guengerich FP: Role of cytochrome P450 enzymes in drug-drug interactions. Adv Pharmacol 43:7–35, 1997

Hachad H, Ragueneau-Majlessi I, Levy RH: New antiepileptic drugs: review on drug interactions. Ther Drug Monit 24(1):91–103, 2002

Hesslinger B, Normann C, Langosch JM, et al: Effects of carbamazepine and valproate on haloperidol plasma levels and on psychopathologic outcome in schizophrenic patients. J Clin Psychopharmacol 19(4):310—315, 1999

Ketter TA, Flockhart DA, Post RM, et al: The emerging role of cytochrome P450 3A in psychopharmacology. J Clin Psychopharmacol 15(6):387–398, 1995

Kudo S, Ishizaki T: Pharmacokinetics of haloperidol: an update. Clin Pharmacokinet 37(6):435–456, 1999

Lucas RA, Gilfillan DJ, Bergstrom RF: A pharmacokinetic interaction between carbamazepine and olanzapine: observations on possible mechanism. Eur J Clin Pharmacol 54(8):639–643, 1998

Martinez C, Albet C, Agundez JA, et al: Comparative in vitro and in vivo inhibition of cytochrome P450 CYP1A2, CYP2D6, and CYP3A by H2-receptor antagonists. Clin Pharmacol Ther 65(4):369–376, 1999

Parker AC, Pritchard P, Preston T, et al: Induction of CYP1A2 activity by carbamazepine in children using the caffeine breath test. Br J Clin Pharmacol 45(2): 176–178, 1998

Sawada Y, Ohtani H: Pharmacokinetics and drug interactions of antidepressive agents (in Japanese). Nippon Rinsho 59(8):1539–1545, 2001

Schrenk D, Brockmeier D, Morike K, et al: A distribution study of CYP1A2 phenotypes among smokers and non-smokers in a cohort of healthy Caucasian volunteers. Eur J Clin Pharmacol 53(5):361–367, 1998

Spina E, Pisani F, Perucca E: Clinically significant pharmacokinetic drug interactions with carbamazepine: an update. Clin Pharmacokinet 31(3):198–214, 1996

von Moltke LL, Greenblatt DJ, Court MH, et al: Inhibition of alprazolam and desipramine hydroxylation in vitro by paroxetine and fluvoxamine: comparison with other selective serotonin reuptake inhibitor antidepressants. J Clin Psychopharmacol 15(2):125–131, 1995

Zevin S, Benowitz NL: Drug interactions with tobacco smoking: an update. Clin Pharmacokinet 36(6):425–438, 1999

Chapter 2

2D6 Case Vignettes

Even though it is not the most important of the P450 enzymes, 2D6 is the usual starting point for any journey into the world of P450, since this was the first P450 enzyme that psychiatrists became familiar with.

The 2D6 enzyme constitutes only 1%–2% of the P450 complement of the liver, yet it is involved in a number of clinically significant drug interactions. Typical substrates include tricyclic antidepressants (TCAs) (Dahl et al. 1993; Sawada and Ohtani 2001), phenothiazines (Dahl-Puustinen et al. 1989; von Bahr et al. 1991), β-blockers (Lennard et al. 1986), and various antiarrhythmic agents (Funck-Bretano et al. 1989; Vandamme et al. 1993). Several of the selective serotonin reuptake inhibitors (SSRIs) are significant inhibitors of 2D6 (Lam et al. 2002), as are TCAs (Lamard et al. 1995; Shin et al. 2002). Quinidine is an extremely potent 2D6 inhibitor (von Moltke et al. 1994).

Fourteen percent of Caucasian individuals are poor metabolizers/polymorphic for 2D6, while 1%–3% of all individuals are ultrarapid metabolizers (Eichelbaum and Evert 1996). There are no verified inducers of 2D6.

References

Dahl ML, Iselius L, Alm C, et al: Polymorphic 2-hydroxylation of desipramine: a population and family study. Eur J Clin Pharmacol 44(5):445–450, 1993

Dahl-Puustinen ML, Liden A, Alm C, et al: Disposition of perphenazine is related to polymorphic debrisoquin hydroxylation in human beings. Clin Pharmacol Ther 46(1):78–81, 1989

Eichelbaum M, Evert B: Influence of pharmacogenetics on drug disposition and response. Clin Exp Pharmacol Physiol 23(10–11):983–985, 1996

Funck-Bretano C, Turgeon J, Woosley RL, et al: Effect of low dose quinidine on encainide pharmacokinetics and pharmacodynamics: influence of genetic polymorphism. J Pharmacol Exp Ther 249(1):134–142, 1989

Lam YW, Gaedigk A, Ereshefsky L, et al: CYP2D6 inhibition by selective serotonin reuptake inhibitors: analysis of achievable steady-state plasma concentrations and the effect of ultrarapid metabolism at CYP2D6. Pharmacotherapy 22(8):1001–1006, 2002

Lamard L, Perault MC, Bouquet S, et al: Cytochrome p450 IID6, its role in psychopharmacology (in French). Ann Med Psychol (Paris) 153(2):140–143, 1995

Lennard MS, Tucker GT, Silas JH, et al: Debrisoquine polymorphism and the metabolism and action of metoprolol, timolol, propranolol and atenolol. Xenobiotica 16(5):435–447, 1986

Sawada Y, Ohtani H: Pharmacokinetics and drug interactions of antidepressive agents (in Japanese). Nippon Rinsho 59(8):1539–1545, 2001

Shin JG, Park JY, Kim MJ, et al: Inhibitory effects of tricyclic antidepressants (TCAs) on human cytochrome P450 enzymes in vitro: mechanism of drug interaction between TCAs and phenytoin. Drug Metab Dispos 30(10):1102–1107, 2002

Vandamme N, Broly F, Libersa C, et al: Stereoselective hydroxylation of mexiletine in human liver microsomes: implication of P450IID6—a preliminary report. J Cardiovasc Pharmacol 21(1):77–83, 1993

von Bahr C, Movin G, Nordin C, et al: Plasma levels of thioridazine and metabolites are influenced by the debrisoquin hydroxylation phenotype. Clin Pharmacol Ther 49(3):234–240, 1991

von Moltke LL, Greenblatt DJ, Cotreau-Bibbo MM, et al: Inhibition of desipramine hydroxylation in vitro by serotonin-reuptake-inhibitor antidepressants, and by quinidine and ketoconazole: a model system to predict drug interactions in vivo. J Pharmacol Exp Ther 268(3):1278–1283, 1994

Palpitations

A 55-year-old woman with refractory major depressive disorder had been maintained on nortriptyline (Pamelor), 100 mg/day, for 1 month, generating a level of 90 ng/mL. Her psychiatrist was especially concerned about her deteriorating condition, so she treated this patient more aggressively than was typical for her. Her concern led her to add paroxetine (Paxil), 20 mg/day. After 2 weeks, the patient had not yet responded, so the paroxetine was further increased to 40 mg/day. Four weeks later, the patient was seen in consultation, at which time she remained depressed and presented with the additional symptoms of dizziness and palpitations. A nortriptyline level drawn several days earlier was 359 ng/mL. An electrocardiogram (ECG) showed sinus tachycardia (114 beats per minute) with no evidence of conduction blockade (N.B. Sandson, self-report, August 1994).

Discussion

This is an example of an inhibitor added to a substrate.

Paroxetine is a potent inhibitor of 2D6 (von Moltke et al. 1995). All TCAs, including nortriptyline, are at least partial substrates of 2D6 (Sawada and Ohtani 2001). Secondary-amine TCAs (e.g., nortriptyline, desipramine, protriptyline) are preferentially metabolized by 2D6 and secondarily metabolized by 3A4 and 1A2 (Dahl et al. 1993; Venkatakrishnan et al. 1999; Zhang and Kaminsky 1995). The metabolism of tertiary-amine TCAs (e.g., imipramine, amitriptyline, clomipramine) is much more complex. The enzymes 2D6, 3A4, 2C19, and 1A2 all play potentially significant roles in the metabolism of tertiary-amine TCAs, and the specifics vary with each individual tertiary-amine TCA (Nielsen et al. 1996; Sawada and Ohtani 2001; Venkatakrishnan et al. 1998; Yang et al. 1999). Prior to adding the paroxetine, a steady-state equilibrium had been established between the stable dosage of nortriptyline and the baseline activity of 2D6 in this individual. With the addition of the paroxetine, the ability of 2D6 to metabolize the nortriptyline was significantly impaired, resulting in a fourfold increase in the nortriptyline blood level, even though there had been no change in the nortriptyline dosage. This led to a state of mild TCA toxicity, manifested by moderate sinus tachycardia.

References

Dahl ML, Iselius L, Alm C, et al: Polymorphic 2-hydroxylation of desipramine: a population and family study. Eur J Clin Pharmacol 44(5):445–450, 1993

Nielsen KK, Flinois JP, Beaune P, et al: The biotransformation of clomipramine in vitro: identification of the cytochrome P450s responsible for the separate metabolic pathways. J Pharmacol Exp Ther 277(3):1659–1664, 1996

Sawada Y, Ohtani H: Pharmacokinetics and drug interactions of antidepressive agents (in Japanese). Nippon Rinsho 59(8):1539–1545, 2001

Venkatakrishnan K, Greenblatt DJ, von Moltke LL, et al: Five distinct human cytochromes mediate amitriptyline N-demethylation in vitro: dominance of CYP 2C19 and 3A4. J Clin Pharmacol 38(2):112–121, 1998

Venkatakrishnan K, von Moltke LL, Greenblatt DJ: Nortriptyline E-10-hydroxylation in vitro is mediated by human CYP2D6 (high affinity) and CYP3A4 (low affinity): implications for interactions with enzyme-inducing drugs. J Clin Pharmacol 39(6):567–577, 1999

von Moltke LL, Greenblatt DJ, Court MH, et al: Inhibition of alprazolam and desipramine hydroxylation in vitro by paroxetine and fluvoxamine: comparison with other selective serotonin reuptake inhibitor antidepressants. J Clin Psychopharmacol 15(2):125–131, 1995

Yang TJ, Krausz KW, Sai Y, et al: Eight inhibitory monoclonal antibodies define the role of individual P450s in human liver microsomal diazepam, 7-ethoxycoumarin, and imipramine metabolism. Drug Metab Dispos 27(1):102–109, 1999

Zhang ZY, Kaminsky LS: Characterization of human cytochromes P450 involved in theophylline 8-hydroxylation. Biochem Pharmacol 50(2):205–211, 1995

Postpartum Delirium

A 32-year-old woman had been previously treated for her depression with desipramine (Norpramin), 125 mg/day, generating a blood level of 142 ng/mL. Although the medication at this level was effective in treating her depression, she experienced unpleasant side effects, such as blurry vision and constipation, despite the fact that desipramine is the least anticholinergic of the TCAs. To avoid these side effects, the patient and her psychiatrist embarked on a successful crossover titration to paroxetine (Paxil), 20 mg/day, which she tolerated without any difficulties. She remained stable while taking the paroxetine for 3 years, and then she had her first child. She continued taking the paroxetine throughout her pregnancy and the postpartum period, but she experienced a severe recurrence of depression. At 10 weeks postpartum, the decision was made to add the desipramine back to her regimen in the hope that this previously effective medication would prove helpful again. It was also hoped that desipramine's noradrenergic reuptake blockade would synergize with paroxetine's SSRI activity. Since desipramine at a dosage of 125 mg/day had proven effective in the past, her psychiatrist rapidly titrated the drug back to that dosage. Within 1 week, the patient became acutely confused and drowsy. A desipramine level was drawn, and both antidepressants were discontinued. The desipramine level was 429 ng/mL.

Discussion

This is an example of a substrate added to an inhibitor.

As a secondary-amine TCA, desipramine is a 2D6 substrate (Dahl et al. 1993), and paroxetine is a strong 2D6 inhibitor (von Moltke et al. 1995). Although this interaction is quite similar to the paroxetine-nortriptyline interaction in the previous case (see "Palpitations," p. 10), this case illustrates the specific dangers of adding a substrate to an inhibitor as opposed to adding an inhibitor to a substrate (as described in "Palpitations"). When a substrate is added to an inhibitor, difficulties can arise when the clinician adheres to preset dosing guidelines, rather than titrating to clinical effect or blood levels. In this case, the decision to titrate to 125 mg/day of desipramine, without taking the "new" presence of paroxetine into account, led to a toxic blood level of desipramine, since the paroxetine was significantly impairing the ability of 2D6 to metabolize the desipramine. This led to the patient's delirious state, as well as an avoidable setback in her treatment.

References

Dahl ML, Iselius L, Alm C, et al: Polymorphic 2-hydroxylation of desipramine: a population and family study. Eur J Clin Pharmacol 44(5):445–450, 1993

von Moltke LL, Greenblatt DJ, Court MH, et al: Inhibition of alprazolam and desipramine hydroxylation in vitro by paroxetine and fluvoxamine: comparison with other selective serotonin reuptake inhibitor antidepressants. J Clin Psychopharmacol 15(2):125–131, 1995

Complicated Crossover

A 35-year-old depressed and anxious man had not responded to an adequate trial of fluoxetine (Prozac), titrated upward to a dosage of 80 mg/day, and the decision was made to switch him to nefazodone (Serzone). This switch was done in "crossover" fashion; the dosage of fluoxetine was initially decreased to 60 mg/day, and the nefazodone was started at a dose of 100 mg bid. After 1 week, the dosage of fluoxetine was further decreased to 40 mg/day and the dosage of nefazodone was increased to 300 mg/day. Within another 3–4 days, the patient reported experiencing dysphoria and severe anxiety, verging on panic.

Discussion

This is an example of an inhibitor added to a substrate.

Fluoxetine is a potent inhibitor of 2D6 (Stevens and Wrighton 1993). One of nefazodone's three principal metabolites is *m*-chlorophenylpiperazine (mCPP), a partial serotonergic agonist (Hamik and Peroutka 1989). Acutely, mCPP has both anxiogenic and dysphoria-producing properties (Ayd 1995). mCPP is principally metabolized by 2D6 (von Moltke et al. 1999). In individuals with normal hepatic function, the baseline level of 2D6 activity is sufficient to prevent clinically significant mCPP accumulation. However, when a medication that liberates physiologically significant amounts of mCPP is coadministered with a strong inhibitor of 2D6 activity, mCPP's metabolism is slowed to the point that symptomatic accumulations of mCPP can occur, as in this case.

References

Ayd F: Lexicon of Psychiatry, Neurology, and the Neurosciences. Baltimore, MD, WIlliams & Wilkins, 1995, p 403

Hamik A, Peroutka SJ: 1-(m-Chlorophenyl)piperazine (mCPP) interactions with neurotransmitter receptors in the human brain. Biol Psychiatry 25(5):569–575, 1989

Stevens JC, Wrighton SA: Interaction of the enantiomers of fluoxetine and norfluoxetine with human liver cytochromes P450. J Pharmacol Exp Ther 266(2): 964–971, 1993

von Moltke LL, Greenblatt DJ, Granda BW, et al: Nefazodone, meta-chlorophenylpiperazine, and their metabolites in vitro: cytochromes mediating transformation, and P450-3A4 inhibitory actions. Psychopharmacology (Berl) 145(1): 113–122, 1999

The Lethargic Librarian

A 55-year-old woman had been given the comorbid diagnoses of bipolar I disorder and obsessive-compulsive disorder (OCD). For the past 5 years, her bipolar disorder had been well controlled with lithium, at a dosage of 900 mg qhs (blood level=0.8 mEq/L). She initially developed an intention tremor from the lithium, but this effect responded well to propranolol (Inderal), 20 mg tid. She since experienced occasional lightheadedness when quickly rising to a standing position, but this side effect of the propranolol was generally well tolerated. In a successful effort to minimize antidepressant use, a trial of cognitive-behavioral therapy was undertaken to manage her OCD symptoms, which responded well to this treatment.

Six months ago, she was laid off from her job as a librarian, which quickly led to a marked worsening of her OCD symptoms. After careful deliberation with her psychiatrist, sertraline (Zoloft) was added to her regimen. A gradual dose titration of sertraline ensued at a rate of 50 mg/month. One week ago, she reached a dosage of 200 mg/day. Her OCD symptoms thus far failed to respond. She was also experienced increasing fatigue, anergy, and dizziness when standing over the past 6 weeks. Yesterday, when standing quickly to answer a ringing phone, she abruptly fell and almost lost consciousness. The responding ambulance crew recorded her pulse at 45 beats per minute, and her blood pressure was 75/40 mm Hg.

Discussion

This is an example of an inhibitor added to a substrate.

Although sertraline is considered a less potent inhibitor of 2D6 than fluoxetine or paroxetine, this is actually a dose-dependent issue. Sertraline, at a dosage of 50–100 mg/day, is indeed a less potent 2D6 inhibitor than 20 mg/day of either fluoxetine or paroxetine. However, 150–200 mg/day of sertraline produces comparable 2D6 inhibition (Alderman et al. 1997; Solai et al. 1997).

Propranolol is a 2D6 substrate (Lennard et al. 1986). When sertraline was titrated to its higher dose range, it impaired the ability of 2D6 to efficiently metabolize the propranolol. This led to an increase in the blood level of propranolol, even though the propranolol dosage had not been changed, which led to worsening orthostatic hypotension, anergy, and the patient's eventual fall.

References

Alderman J, Preskorn SH, Greenblatt DJ, et al: Desipramine pharmacokinetics when coadministered with paroxetine or sertraline in extensive metabolizers. J Clin Psychopharmacol 17(4):284–291, 1997

Lennard MS, Tucker GT, Silas JH, et al: Debrisoquine polymorphism and the metabolism and action of metoprolol, timolol, propranolol and atenolol. Xenobiotica 16(5):435–447, 1986

Solai LK, Mulsant BH, Pollock BG, et al: Effect of sertraline on plasma nortriptyline levels in depressed elderly. J Clin Psychiatry 58(10):440–443, 1997

Panicked and Confused

A 35-year-old man with long-standing schizoaffective disorder, bipolar type, and alcohol dependence in full remission was being stably maintained on haloperidol (Haldol), 10 mg qhs, and divalproex sodium (Depakote), 1,000 mg bid. Benztropine (Cogentin), 2 mg bid, alleviated his haloperidol-induced tremor and stiffness without causing any further side effects. Over the past 2 years, both of his parents died from medical causes; this led to the emergence of frequent and debilitating panic attacks. His psychiatrist hoped to alleviate these panic attacks by adding paroxetine (Paxil), 20 mg qhs, to the patient's regimen—declining to use benzodiazepines, to avoid rekindling his alcohol use. Within 5 days, the patient experienced new-onset blurring of his vision, urinary retention, and mild memory impairment. After taking a nap and waking in the early evening, he could not remember what day it was or whether it was morning or evening, thus inducing another severe panic attack. The psychiatrist told the patient to stop taking the paroxetine and sent him to have blood levels drawn for his medications. His haloperidol and divalproex levels were essentially unchanged and in the normal range, but his benztropine level (no baseline) was 42 ng/mL (levels>25 ng/mL are considered toxic) (Specialty Laboratories 2001). On receiving this result, the psychiatrist held the patient's benztropine, and his anticholinergic symptoms abated over the next 3 days.

Discussion

This is an example of an inhibitor added to a substrate.

Benztropine is believed to be a 2D6 substrate, and there have been several documented instances that suggest that 2D6-inhibiting SSRIs, such as paroxetine (von Moltke et al. 1995), inhibit benztropine's metabolism (Armstrong and Schweitzer 1997). The addition of paroxetine to the regimen impaired the ability of 2D6 to efficiently metabolize the benztropine, which led to an increase in the blood level of benztropine, even though the benztropine dosage had not been changed. The increased benztropine level led to the emergence of several anticholinergic symptoms (blurry vision due to mydriasis, urinary retention, and mild confusion).

References

Armstrong SC, Schweitzer SM: Delirium associated with paroxetine and benztropine combination. Am J Psychiatry 154(4):581–582, 1997

Specialty Laboratories: Directory of Services, 2001. Santa Monica, CA, Specialty Laboratories, 2001, p 55

von Moltke LL, Greenblatt DJ, Court MH, et al: Inhibition of alprazolam and desipramine hydroxylation in vitro by paroxetine and fluvoxamine: comparison with other selective serotonin reuptake inhibitor antidepressants. J Clin Psychopharmacol 15(2):125–131, 1995

Cranky and Crampy

An 86-year-old man with Alzheimer's dementia was experiencing increasing difficulties living independently. His family orchestrated his transition to an assisted-living facility. They also obtained a consultation with a psychiatrist, who started the patient on donepezil (Aricept) and titrated his dosage to 10 mg/day. Vitamin E was also added and titrated to a dosage of 2,000 IU/day. His internist had previously prescribed diltiazem (Cardizem), which the patient was taking on a chronic basis for control of his hypertension. Over the next 2 months, he was noted by facility staff to be more cranky and irritable. One day, he became so frustrated about receiving turkey for lunch rather than roast beef that he hurled his lunch tray across his room. The psychiatrist was reconsulted. In the interview with the patient, the patient revealed his resentment, hopelessness, and sense of abandonment by his family, as well as decreased sleep and appetite. The psychiatrist opted to start the patient on paroxetine (Paxil), 10 mg/day for 3 days and 20 mg/day thereafter. In the week after starting the paroxetine, the patient experienced increasingly distressing abdominal cramping and diarrhea. The psychiatrist then stopped the paroxetine, waited 2 weeks, and initiated a trial of citalopram (Celexa), which the patient tolerated without difficulty.

Discussion

This is an example of an inhibitor added to a substrate.

Donepezil is metabolized primarily by 2D6 and secondarily by 3A4 (Aricept [package insert] 2000). The addition of paroxetine potently inhibited 2D6 (von Moltke et al. 1995), thus impairing its ability to efficiently metabolize the donepezil. The presence of diltiazem, a potent 3A4 inhibitor (Sutton et al. 1997), prevented 3A4 from serving as an effective accessory metabolic pathway for donepezil. Thus, the addition of paroxetine increased the blood level of donepezil, an acetylcholinesterase inhibitor, which led to the cholinergic side effects of cramping and diarrhea.

References

Aricept (package insert). Teaneck, NJ, Eisai Inc, December 2000

Sutton D, Butler AM, Nadin L, et al: Role of CYP3A4 in human hepatic diltiazem N-demethylation: inhibition of CYP3A4 activity by oxidized diltiazem metabolites. J Pharmacol Exp Ther 282(1):294–300, 1997

von Moltke LL, Greenblatt DJ, Court MH, et al: Inhibition of alprazolam and desipramine hydroxylation in vitro by paroxetine and fluvoxamine: comparison with other selective serotonin reuptake inhibitor antidepressants. J Clin Psychopharmacol 15(2):125–131, 1995

The Potential Perils of Frugality

A 19-year-old man with newly diagnosed OCD had experienced only a marginal response to paroxetine (Paxil), titrated to a dosage of 60 mg/day. He would routinely spend 4 or more hours each day cleaning himself and straightening his apartment. He was able to maintain gainful employment only by surviving on 3–4 hours of sleep each night, leading him to a state of exhaustion by each Friday. The decision was made to add clomipramine (Anafranil) at an initial dosage of 25 mg/day. After 4 days, the dosage was increased to 50 mg/day. A clomipramine blood level (reported as clomipramine+desmethylclomipramine) drawn 1 week later was in the normal range, and he tolerated this addition without significant side effects. Over the next 6 weeks, the patient began to experience significant improvement in his OCD symptoms. The patient did well on this regimen for 2 years. However, he then lost his job during a flurry of downsizing, which led to the loss of his insurance coverage. In view of his financial difficulties, the patient and his psychiatrist decided to try the free samples of citalopram (Celexa) that were available at the clinic in the place of the paroxetine. No side effects emerged, but within 1 month the patient's cleaning rituals were significantly worse, although not as severe as they had been before the addition of the clomipramine.

Discussion

This is an example of reversal of inhibition.

Paroxetine is a potent 2D6 inhibitor (von Moltke et al. 1995). Clomipramine is a tertiary-amine TCA whose metabolism depends most on the intact functioning of 2C19, 3A4, and 2D6, with 1A2 serving as a secondary enzyme (Nielsen et al. 1996). As such, paroxetine's inhibition of 2D6 was able to significantly impair the metabolism of both clomipramine and desmethylclomipramine, leading to an increase in this total blood level. (Desmethylclomipramine is clomipramine's primary metabolite via demethylation by 2C19 and 3A4. 2D6 performs subsequent hydroxylation.) Thus, the presence of paroxetine enabled a relatively low dosage of clomipramine to produce a therapeutic response. Citalopram, however, lacks significant inhibitory activity at any of the P450 enzymes (Brosen and Naranjo 2001). Thus, with the replacement of paroxetine by citalopram, clomipramine's metabolism was no longer being inhibited, with the result that the clomipramine+desmethylclomipramine level declined. This led to a significant loss of clinical efficacy in this case.

References

Brosen K, Naranjo CA: Review of pharmacokinetic and pharmacodynamic interaction studies with citalopram. Eur Neuropsychopharmacol 11(4):275–283, 2001
Nielsen KK, Flinois JP, Beaune P, et al: The biotransformation of clomipramine in vitro: identification of the cytochrome P450s responsible for the separate metabolic pathways. J Pharmacol Exp Ther 277(3):1659–1664, 1996
von Moltke LL, Greenblatt DJ, Court MH, et al: Inhibition of alprazolam and desipramine hydroxylation in vitro by paroxetine and fluvoxamine: comparison with other selective serotonin reuptake inhibitor antidepressants. J Clin Psychopharmacol 15(2):125–131, 1995

The Cure Can Be Worse Than the Disease

A 35-year-old man with a history of recurrent melancholic major depressive disorder had been stably maintained for over 3 years on a regimen of desipramine (Norpramin), 200 mg/day, with a stable blood level of around 150 ng/mL. The patient was also a two-pack-per-day smoker. His father had died in his mid-40s from a myocardial infarction. Accordingly, this patient's family physician strongly advised the patient to participate in an aggressive smoking cessation program. As part of this program, the physician started the patient on bupropion (Zyban, in this application), titrating to a dose of 150 mg bid. One week after reaching this dose, the patient began to experience frequent palpitations and accompanying lightheadedness. He reported these symptoms to his family physician, who evaluated the patient. His ECG showed a sinus tachycardia of 132 beats per minute and QRS interval widening to a duration of 160 msec. The patient was admitted to a telemetry unit, where a desipramine level was drawn before the drug was discontinued, and the result was 623 ng/mL. A week after the desipramine was stopped, his symptoms fully remitted.

Discussion

This is an example of an inhibitor added to a substrate.

Desipramine, as a secondary-amine TCA, is metabolized primarily at 2D6 (Dahl et al. 1993). Bupropion is a relatively potent 2D6 inhibitor, capable of elevating desipramine blood levels to two to five times baseline (Wellbutrin [package insert] 2001). Thus, the addition of bupropion (albeit in the form of Zyban) impaired the ability of 2D6 to efficiently metabolize the desipramine, which led to a more than fourfold increase in the desipramine blood level in this case, even though the desipramine dosage had not been changed. The resultant cardiac toxicity (via excessive quinidine-like effects of toxic TCA levels) led to the symptoms of palpitations and dizziness.

References

Dahl ML, Iselius L, Alm C, et al: Polymorphic 2-hydroxylation of desipramine: a population and family study. Eur J Clin Pharmacol 44(5):445–450, 1993

Wellbutrin (package insert). Research Triangle Park, NC, GlaxoSmithKline, 2001

Anergy Through Synergy

A 27-year-old woman failed to respond to numerous sequential antidepressant monotherapy trials, including trials of several SSRIs, nefazodone (Serzone), venlafaxine (Effexor), and mirtazapine (Remeron). Her psychiatrist opted to try amitriptyline (Elavil), titrated to a dosage of 200 mg/day. At this dosage, her blood level of amitriptyline + nortriptyline (amitriptyline's primary metabolite) was 185 ng/mL. She remained on this dosage for 6 weeks, without discernible improvement. As an attempt at augmentation, the psychiatrist added bupropion (Wellbutrin SR), titrating to a dose of 200 mg bid. Over the next 10 days, the patient experienced increasing lethargy, dry mouth, and blurry vision. This prompted the psychiatrist to recheck the amitriptyline + nortriptyline blood level, as well as to obtain an ECG. The ECG revealed a sinus tachycardia of 128 beats per minute with a change in QTc interval from 436 msec to 509 msec. Her amitriptyline+nortriptyline blood level was 732 ng/mL (M. Levinson, personal communication, May 2002).

Discussion

This is an example of an inhibitor added to a substrate.

Bupropion is a relatively potent 2D6 inhibitor (Wellbutrin [package insert] 2001). Amitriptyline is a tertiary-amine TCA whose metabolism depends most on the intact functioning of 2C19, 3A4, and 2D6, with 1A2 serving as a secondary enzyme (Venkatakrishnan et al. 1998). Nortriptyline is amitriptyline's primary metabolite via demethylation by 2C19 and 3A4. 2D6 performs subsequent hydroxylation. Thus, the addition of bupropion significantly impaired the ability of 2D6 to contribute to the metabolism of amitriptyline and nortriptyline, which led to an almost fourfold increase in the amitriptyline+nortriptyline blood level. This state of TCA toxicity caused the changes in cardiac conduction described in the case. A similar effect has been observed when bupropion has been added to imipramine (Tofranil), resulting in elevations of imipramine+desipramine blood levels (J.R. Oesterheld, personal communication, May 2002).

References

Venkatakrishnan K, Greenblatt DJ, von Moltke LL, et al: Five distinct human cytochromes mediate amitriptyline N-demethylation in vitro: dominance of CYP 2C19 and 3A4. J Clin Pharmacol 38(2):112–121, 1998

Wellbutrin (package insert). Research Triangle Park, NC, GlaxoSmithKline, 2001

Migraine Misery

A 65-year-old woman with long-standing migraines had failed multiple treatment regimens—verapamil (Calan), divalproex sodium (Depakote), indomethacin (Indocin), and diet—and her history of chronic stable angina contraindicated the use of sumatriptan (Imitrex). Her neurologist found that use of tramadol (Ultram) as needed provided reliable relief from the pain. One year later, her cardiac condition deteriorated to the point that she experienced arrhythmias, and her internist prescribed quinidine (Quinaglute). Two weeks later, she had another migraine headache, and she took her usual dose of tramadol, but this time she did not experience any significant pain relief.

Discussion

This is an example of a substrate added to an inhibitor.

Tramadol is a 2D6 substrate, as well as a pro-drug. Tramadol itself is probably not an effective analgesic compound, and it relies on conversion to M1 by 2D6 to achieve full analgesic potency (Sindrup et al. 1999). Quinidine is an extremely potent 2D6 inhibitor (von Moltke et al. 1994). Thus, when the usual analgesic dose of tramadol was given to this patient, whose complement of 2D6 had been rendered much less active by the quinidine, tramadol's transformation to M1 was inhibited, resulting in a loss of analgesic efficacy (Poulsen et al. 1996).

This interaction is somewhat counterintuitive, in that substrate-inhibitor interactions, when problematic, usually result in substrate toxicity states. However, when the substrate is a pro-drug, the substrate-inhibitor interaction may lead to a lack of efficacy when the pro-drug is not transformed into the desired active metabolite.

References

Poulsen L, Arendt-Nielsen L, Brosen K, et al: The hypoalgesic effect of tramadol in relation to CYP2D6. Clin Pharmacol Ther 60(6):636–644, 1996

Sindrup SH, Madsen C, Brosen K, et al: The effect of tramadol in painful polyneuropathy in relation to serum drug and metabolite levels. Clin Pharmacol Ther 66(6):636–641, 1999

von Moltke LL, Greenblatt DJ, Cotreau-Bibbo MM, et al: Inhibition of desipramine hydroxylation in vitro by serotonin-reuptake-inhibitor antidepressants, and by quinidine and ketoconazole: a model system to predict drug interactions in vivo. J Pharmacol Exp Ther 268(3):1278–1283, 1994

Sad and Sore

A 35-year-old woman underwent a splenectomy following a car accident in which she experienced severe blunt abdominal trauma and multiple other orthopedic injuries. Her surgeon prescribed Vicodin (acetaminophen + hydrocodone) for postoperative pain, which provided adequate analgesia. The accident led to a recurrence of her major depressive disorder, for which she renewed her use of fluoxetine (Prozac), 20 mg/day. Over the next 3 weeks, she began to experience a remission of her depressive symptoms, but she also experienced a corresponding waning of analgesic effect from her usual dose of Vicodin. Her psychiatrist discontinued her fluoxetine, and 4 days later started her on citalopram (Celexa), 20 mg/day. Over the next 3 weeks, she regained the analgesic response from the Vicodin, with no loss of antidepressant efficacy (S.C. Armstrong, personal communication, May 2002).

Discussion

This is an example of an inhibitor added to a substrate.

Just as tramadol is a pro-drug for M1 (see “Migraine Misery,” p. 21), hydrocodone is a pro-drug for hydromorphone (Dilaudid). Hydrocodone is a substrate of 2D6, which catalyzes its transformation into hydromorphone (Otton et al. 1993). Fluoxetine is a strong 2D6 inhibitor (Stevens and Wrighton 1993). Thus, the addition of the fluoxetine impaired the ability of 2D6 to transform hydrocodone into hydromorphone, even as it provided antidepressant efficacy. The switch to citalopram, which lacks a significant P450 inhibitory profile (Brosen and Naranjo 2001), led to a cessation of clinically significant 2D6 inhibition. This allowed for the renewed conversion of hydrocodone to hydromorphone and a return of analgesia, with no loss of antidepressant efficacy in this patient.

References

Brosen K, Naranjo CA: Review of pharmacokinetic and pharmacodynamic interaction studies with citalopram. Eur Neuropsychopharmacol 11(4):275–283, 2001

Otton SV, Schadel M, Cheung SW, et al: CYP2D6 phenotype determines the metabolic conversion of hydrocodone to hydromorphone. Clin Pharmacol Ther 54(5):463–472, 1993

Stevens JC, Wrighton SA: Interaction of the enantiomers of fluoxetine and norfluoxetine with human liver cytochromes P450. J Pharmacol Exp Ther 266(2): 964–971, 1993

Fungal Frustration

A 55-year-old man with a history of chronic major depressive disorder was being reasonably well managed with imipramine (Tofranil), 200 mg/day, generating imipramine+desipramine blood levels that ranged from 175 to 225 ng/mL. He developed a painful and persistent case of onychomycosis under the nail of his left big toe. It became so severe that it began to affect his gait, and it did not respond to topical antifungal remedies (such as clotrimazole [Lotrimin] cream). He consulted a dermatologist, who prescribed oral terbinafine (Lamisil), 250 mg/day. Within 1 week, the patient experienced worsening dysphoria and impaired concentration, sleep, and appetite. Suspecting that these apparent neurovegetative changes represented one of the patient's occasional and sporadic illness exacerbations, his psychiatrist increased the dosage of imipramine from 200 mg/day to 225 mg/day. One week later, the patient was additionally experiencing marked muscle twitching, dry mouth, and dizziness so severe that he fell just after rising from a sitting position. At that point, his imipramine+desipramine blood level was drawn and was found to be 575 ng/mL. The imipramine was discontinued and the patient visited a local ER for observation, but he was released the same day since his ECG revealed only mild sinus tachycardia. His terbinafine was continued throughout. After learning that his patient had been prescribed and was taking terbinafine, the psychiatrist eventually had the patient restart imipramine, titrating to a dosage of 100 mg/day, which generated an imipramine+desipramine blood level of 230 ng/mL. After the patient's onychomycosis resolved, the terbinafine was discontinued, whereupon the imipramine was again retitrated to 200 mg/day and his blood level remained basically constant (between 200 and 250 ng/mL) (Teitelbaum and Pearson 2001).

Discussion

This is an example of an inhibitor added to a substrate.

Imipramine is a tertiary-amine TCA whose metabolism depends most on the intact functioning of 2C19, 3A4, 2D6, and 1A2 (Yang et al. 1999). Desipramine is imipramine's primary metabolite via demethylation by 2C19 and 3A4. 2D6 performs subsequent hydroxylation. Terbinafine is a strong 2D6 inhibitor (Abdel-Rahman et al. 1999). Introducing terbinafine into the patient's regimen prevented 2D6 from effectively metabolizing both imipramine and desipramine. This led to a roughly threefold increase in the blood level of imipramine+desipramine, even though the dosage had been increased from 200 mg/day to only 225 mg/day. This TCA toxicity state led to the patient's muscle twitching, dry mouth, and dizziness (which led to a fall), as well as to other side effects (dysphoria, poor appetite, decreased sleep, and impaired concentration) that mimicked an exacerbation of his major depressive disorder.

References

Abdel-Rahman SM, Marcucci K, Boge T, et al: Potent inhibition of cytochrome P-450 2D6–mediated dextromethorphan O-demethylation by terbinafine. Drug Metab Dispos 27(7):770–775, 1999

Teitelbaum ML, Pearson VE: Imipramine toxicity and terbinafine. Am J Psychiatry 158(12):2086, 2001

Yang TJ, Krausz KW, Sai Y, et al: Eight inhibitory monoclonal antibodies define the role of individual P450s in human liver microsomal diazepam, 7-ethoxycoumarin, and imipramine metabolism. Drug Metab Dispos 27(1):102–109, 1999

Bloated

A 37-year-old woman with a long-standing history of panic disorder had been experiencing only one panic attack per month for the past several years while taking amitriptyline (Elavil), 175 mg/day. At this dosage, her blood level of amitriptyline+nortriptyline ranged from 150 to 180 ng/mL. However, her pet Labrador, who had helped provide a sense of companionship and security, finally succumbed to old age and passed away. Following the loss of this cherished pet, her panic attacks increased in frequency (to five times per week) and intensity. In an effort to address this, her psychiatrist added clonazepam (Klonopin), 0.5 mg tid, to her regimen. At the same time, the psychiatrist also added paroxetine (Paxil), 10 mg/day for 4 days and 20 mg/day thereafter, while simultaneously decreasing the amitriptyline to 150 mg/day. Ten days after starting the paroxetine, the patient began to experience abdominal cramping, back pain, and a pervasive feeling of being bloated. She had a significant decrease in her frequency of bowel movements. Her bloating led to increasing nausea and even vomiting. Her increasing inability to tolerate oral intake led her to visit the local ER. She was admitted to the hospital, where a gastrointestinal workup revealed a partial small bowel obstruction due to an adynamic ileus. An amitriptyline+nortriptyline blood level was found to be 694 ng/mL, whereupon she was also placed on a heart monitor (S. Ruths, personal communication, June 2002).

Discussion

This is an example of an inhibitor added to a substrate.

Amitriptyline is a tertiary-amine TCA whose metabolism depends most on the intact functioning of 2C19, 3A4, and 2D6, with 1A2 serving as a secondary enzyme (Venkatakrishnan et al. 1998). Nortriptyline is amitriptyline's primary metabolite via demethylation by 2C19 and 3A4. 2D6 performs subsequent hydroxylation. Paroxetine is a strong 2D6 inhibitor (von Moltke et al. 1995). Thus, the addition of paroxetine significantly impaired the ability of 2D6 to contribute to the metabolism of amitriptyline and nortriptyline, which led to a three- to fourfold increase in the amitriptyline+nortriptyline blood level, despite a modest decrease in amitriptyline dosage (from 175 mg/day to 150 mg/day). The resultant state of TCA toxicity led to the patient's adynamic ileus and associated cramping, bloating, vomiting, and back pain. It also led to the quinidine-like changes in cardiac conduction that were observed after she was placed on a heart monitor.

References

Venkatakrishnan K, Greenblatt DJ, von Moltke LL, et al: Five distinct human cytochromes mediate amitriptyline N-demethylation in vitro: dominance of CYP 2C19 and 3A4. J Clin Pharmacol 38(2):112–121, 1998

von Moltke LL, Greenblatt DJ, Court MH, et al: Inhibition of alprazolam and desipramine hydroxylation in vitro by paroxetine and fluvoxamine: comparison with other selective serotonin reuptake inhibitor antidepressants. J Clin Psychopharmacol 15(2):125–131, 1995

Achy Breaky Heart

A 42-year-old man with chronic schizophrenia had done well for several years while taking thioridazine (Mellaril), 400 mg qhs. He then attempted to fulfill his lifelong dream of recording several country-and-western songs that he had written. He booked time at a recording studio and made a record, but over the ensuing months he was unable to convince any radio disc jockeys to play any of his songs on the air. His lack of success in this regard led to a depressive episode, for which his psychiatrist added paroxetine (Paxil), 20 mg qhs, to his regimen. After 1 week, the patient complained to his psychiatrist about a new-onset resting tremor, stiffness, and a disconcerting feeling that his heart was racing. An ECG showed moderate sinus tachycardia (126 beats per minute) and a QTc interval of 511 msec. The psychiatrist discontinued the paroxetine, and the patient's symptoms abated within another week.

Discussion

This is an example of an inhibitor added to a substrate.

Thioridazine is a 2D6 substrate (von Bahr et al. 1991). Paroxetine is a strong 2D6 inhibitor (von Moltke et al. 1995). The addition of paroxetine to the patient's thioridazine led to a marked reduction in the metabolism of thioridazine by 2D6. This marked reduction in metabolism caused a marked elevation in the blood level of thioridazine, as evidenced by the new-onset symptoms of neuroleptic-induced parkinsonism, even though the dosage of thioridazine had remained constant. The new sensation of a "racing heart" and his new QTc interval of 511 msec resulted from this interaction. The discontinuation of paroxetine led to a cessation of 2D6 inhibition. This allowed the blood level of thioridazine to return to its baseline, which led to a restoration of normal cardiac functioning.

Elevations in thioridazine blood levels and the resultant lengthening of the QTc interval are of great clinical importance, as they can predispose to torsades de pointes, a dangerous cardiac arrhythmia. It is for this reason that the package insert for thioridazine has recently been modified. It now states that thioridazine should not be prescribed with 2D6 inhibitors (Mellaril [package insert] 2000).

References

Mellaril (package insert). East Hanover, NJ, Novartis Pharmaceuticals, June 2000

von Bahr C, Movin G, Nordin C, et al: Plasma levels of thioridazine and metabolites are influenced by the debrisoquin hydroxylation phenotype. Clin Pharmacol Ther 49(3):234–240, 1991

von Moltke LL, Greenblatt DJ, Court MH, et al: Inhibition of alprazolam and desipramine hydroxylation in vitro by paroxetine and fluvoxamine: comparison with other selective serotonin reuptake inhibitor antidepressants. J Clin Psychopharmacol 15(2):125–131, 1995

Postpartum Psychosis

A 35-year-old mother of two children had responded well to a regimen of citalopram (Celexa), 20 mg/day, for her recurrent major depressive disorder. However, following the birth of her second child, her schedule became so hectic that she intermittently forgot to take her citalopram. She and her psychiatrist decided to switch to the once-per-week preparation of fluoxetine (Prozac), and she responded well to a dose of 90 mg/week. One year later, she again became pregnant, and she continued taking her fluoxetine throughout the pregnancy. However, following the birth of this child, she experienced a postpartum depression with prominent psychotic features, specifically nihilistic delusions of worthlessness and auditory hallucinations telling her that "the children will be better off without you." During previous psychotic depressions, she had taken thioridazine (Mellaril), 200 mg/day, a dosage that was both effective and well tolerated in the past. Her psychiatrist admitted her to the local day hospital program and again restarted her on thioridazine, 200 mg/day, per the patient's firm wishes. However, within 5 days, she complained about having a "thick tongue," painful spasms of her thigh muscles, and frequent lightheadedness. The psychiatrist discontinued the thioridazine. The patient required benztropine (Cogentin) to control these symptoms for the next 2 days, after which they abated. Thioridazine was restarted and titrated to a dose of 75 mg qhs, which the patient tolerated without difficulty, and which proved helpful for her nihilistic delusions and associated hallucinations over the next 2 weeks (S.C. Armstrong, personal communication, May 2002).

Discussion

This is an example of a substrate added to an inhibitor.

Thioridazine is a 2D6 substrate (von Bahr et al. 1991), and fluoxetine is a strong 2D6 inhibitor (Stevens and Wrighton 1993). When thioridazine was used in the past with antidepressants that did not inhibit 2D6 (such as citalopram), 200 mg/day was an effective and tolerable dosage. However, when fluoxetine significantly impaired the ability of 2D6 to efficiently metabolize the thioridazine, then 200 mg/day of thioridazine generated a significantly higher blood level (probably by a factor of two to four) because of the reduction in 2D6's ability to metabolize the thioridazine. These higher concentrations of thioridazine then generated the symptoms of acute dystonia and orthostatic hypotension described by the patient.

Although it is not a central issue in this case, the thioridazine package insert has recently been modified. It now states that thioridazine should not be prescribed with 2D6 inhibitors because of the risk of thioridazine blood level increases, predisposing to prolongations of the QTc interval and subsequent arrhythmias (Mellaril [package insert] 2000).

References

Mellaril (package insert). East Hanover, NJ, Novartis Pharmaceuticals, June 2000

Stevens JC, Wrighton SA: Interaction of the enantiomers of fluoxetine and norfluoxetine with human liver cytochromes P450. J Pharmacol Exp Ther 266(2): 964–971, 1993

von Bahr C, Movin G, Nordin C, et al: Plasma levels of thioridazine and metabolites are influenced by the debrisoquin hydroxylation phenotype. Clin Pharmacol Ther 49(3):234–240, 1991

Anagram

A 31-year-old man with schizophrenia had been stably maintained on perphenazine (Trilafon) monotherapy, at a dosage of 32 mg/day, for the past 2 years. Following a breakup with his longtime girlfriend, he entered a period of persisting depressive symptoms (dysphoria, terminal insomnia, poor appetite, poor energy, and anhedonia). During a depressive episode 5 years earlier, he had been taking molindone (Moban), 40 mg/day (equivalent potency to his current regimen of perphenazine) because of concerns about weight, and his antidepressant had been imipramine (Tofranil), 175 mg/day, which generated imipramine+ desipramine blood levels ranging from 225 ng/mL to 275 ng/mL. When the patient discussed antidepressant selection with his psychiatrist, the patient insisted on once again taking imipramine. His reasoning was twofold. First, it had helped in the past. Second, he endorsed the magical belief that these medications, imipramine and perphenazine, were "meant" to go together, since their trade names (Tofranil and Trilafon) had all the same letters. His psychiatrist titrated the imipramine to the same dosage as previously (175 mg/day). Within 1 week, the patient experienced a cluster of new symptoms, including jitteriness, cogwheel rigidity, dry mouth, blurry vision, urinary hesitancy, and dizziness. The psychiatrist discontinued the imipramine and ordered an imipramine+desipramine blood level, which was 458 ng/mL.

Discussion

This is a combined example of a 2D6 inhibitor (imipramine) added to a 2D6 substrate (perphenazine) and a 2D6 substrate (imipramine) added to a 2D6 inhibitor (perphenazine).[1]

When the imipramine was added to the perphenazine, it prevented 2D6 from effectively metabolizing the perphenazine. This led to an increase in the blood level of perphenazine despite constant dosing, resulting in new-onset akathisia and cogwheeling. Conversely, since perphenazine is also a 2D6 inhibitor, this prevented the efficient metabolism of imipramine. Imipramine was added at a dosage that was appropriate in the absence of a 2D6 inhibitor (molindone does not significantly inhibit 2D6). However, since the perphenazine was present, imipramine's metabolism was significantly inhibited, and the resulting imipramine+desipramine blood level was roughly double what had been expected by the psychiatrist, leading to the patient's anticholinergic symptoms (dry mouth, blurry vision, urinary hesitancy) and orthostasis-induced complaint of dizziness (Brosen et al. 1986).

Imipramine has been cited in a number of other cases in this chapter, but only as a substrate of 2D6, not as an inhibitor. Clearly imipramine did not cease to be an active 2D6 inhibitor in those other examples, but there were no other 2D6 substrates that imipramine could significantly inhibit in those other cases that would have illustrated this property of imipramine. Indeed, most TCAs have significant 2D6 inhibitory capabilities (Lamard et al. 1995; Shin et al. 2002).

References

Brosen K, Gram LF, Klysner R, et al: Steady-state levels of imipramine and its metabolites: significance of dose-dependent kinetics. Eur J Clin Pharmacol 30(1):43–49, 1986

Dahl-Puustinen ML, Liden A, Alm C, et al: Disposition of perphenazine is related to polymorphic debrisoquin hydroxylation in human beings. Clin Pharmacol Ther 46(1):78–81, 1989

---

[1]Dahl-Puustinen et al. 1989; Gex-Fabry et al. 2001; Sawada and Ohtani 2001; Shin et al. 2002; von Bahr et al. 1991.

Gex-Fabry M, Balant-Gorgia A, Balant L: Therapeutic drug monitoring databases for postmarketing surveillance of drug-drug interactions. Drug Saf 24(13): 947–959, 2001

Lamard L, Perault MC, Bouquet S, et al: Cytochrome p450 IID6: its role in psychopharmacology (in French). Ann Med Psychol (Paris) 153(2):140–143, 1995

Sawada Y, Ohtani H: Pharmacokinetics and drug interactions of antidepressive agents (in Japanese). Nippon Rinsho 59(8):1539–1545, 2001

Shin JG, Park JY, Kim MJ, et al: Inhibitory effects of tricyclic antidepressants (TCAs) on human cytochrome P450 enzymes in vitro: mechanism of drug interaction between TCAs and phenytoin. Drug Metab Dispos 30(10):1102–1107, 2002

von Bahr C, Movin G, Nordin C, et al: Plasma levels of thioridazine and metabolites are influenced by the debrisoquin hydroxylation phenotype. Clin Pharmacol Ther 49(3):234–240, 1991

Dry Heat

A 27-year-old man with polysubstance dependence had discontinued his use of methadone (Dolophine), 60 mg/day, and had again begun abusing speedball (intravenous heroin and cocaine). After being confronted by several family members, he chose to present for treatment at a local dual-diagnosis day hospital. The psychiatrist there diagnosed him with depressive disorder not otherwise specified, and a past history revealed prior depressive episodes with a good past response to desipramine (Norpramin). The psychiatrist started him on desipramine, titrating the dosage to 200 mg/day, which generated a desipramine blood level of 154 ng/mL. The patient tolerated the desipramine quite well, with no significant side effects. Soon before his transition to routine outpatient care, his methadone was restarted at the prior dosage of 60 mg/day. Over the ensuing week, he experienced increasing sedation and dry mouth. Also, this was a very hot time of the summer, and he found that despite feeling especially hot, he was sweating very little. He reported these symptoms and his rapidly increasing lethargy to his psychiatrist. The psychiatrist found that the patient had a temperature of 101°, whereupon he discontinued the desipramine and ordered a stat ECG and desipramine blood level. The ECG showed sinus tachycardia of 110 beats per minute and a desipramine blood level of 393 ng/mL.

Discussion

This is an example of an inhibitor added to a substrate.

Desipramine is primarily a substrate of 2D6 (Dahl et al. 1993), as are other secondary-amine TCAs. Methadone is a moderately strong inhibitor of

2D6 (as well as 3A4) (Wu et al. 1993). The introduction of methadone significantly impaired the ability of 2D6 to metabolize the desipramine, thus leading to a significant increase in the desipramine blood level (Maany et al. 1989), even though the dosage had not been further increased. This toxic desipramine level resulted in the patient's tachycardia, sedation, dry mouth, and more general anhydrosis, which predisposed him to develop a fever and severe lethargy during a hot summer week.

It is worth noting that whereas the blood levels of tertiary-amine TCAs (imipramine, clomipramine, amitriptyline, and doxepin) may be elevated by the addition of either 2D6 or 3A4 inhibitors, the blood levels of secondary-amine TCAs (nortriptyline, desipramine, and protriptyline) will generally be increased by the addition of 2D6 inhibitors, but not 3A4 inhibitors. This is because the hydroxylation performed by 2D6 is quite important to the metabolism of both secondary- and tertiary-amine TCAs, whereas the demethylation performed by 3A4 is much less crucial to the metabolism of secondary-amine TCAs.

References

Dahl ML, Iselius L, Alm C, et al: Polymorphic 2-hydroxylation of desipramine: a population and family study. Eur J Clin Pharmacol 44(5):445–450, 1993

Maany I, Dhopesh V, Arndt IO, et al: Increase in desipramine serum levels associated with methadone treatment. Am J Psychiatry 146(2):1611–1613, 1989

Wu D, Otton SV, Sproule BA, et al: Inhibition of human cytochrome P450 2D6 (CYP2D6) by methadone. Br J Clin Psychopharmacol 35(1):30–34, 1993

Window

A 42-year-old man with recurrent major depression and a remote history of intravenous drug abuse was being stably maintained on nortriptyline (Pamelor), 100 mg/day (most recent blood level=91 ng/mL). A new girlfriend insisted on his obtaining an HIV test before they began to have sexual relations, whereupon he discovered that he was HIV-positive. After verifying that this result was not a false positive, his internist prescribed ritonavir (Norvir). Over the next 2 weeks, the patient experienced increasing irritability, dysphoria, insomnia, and demoralization. Fearing that a drug interaction had occurred, the internist ordered a nortriptyline blood level, which had risen to 218 ng/mL. The internist decreased the nortriptyline dosage to 50 mg/day, and another blood level 1 week later was 112 ng/mL. After another week, the depressive symptoms had remitted (K.L. Cozza, personal communication, May 2002).

Discussion

This is an example of an inhibitor added to a substrate.

Nortriptyline is a 2D6 substrate (Sawada and Ohtani 2001), and ritonavir is a 2D6 inhibitor (as well as an inhibitor of most P450 enzymes except 1A2, and a later inducer of 3A4 only) (von Moltke et al. 1998). The addition of the ritonavir impaired the ability of 2D6 to efficiently metabolize the nortriptyline, thus yielding a higher blood level of nortriptyline even though the dosage had not been increased. Studies have suggested that nortriptyline demonstrates a therapeutic blood level window of 50 to 150 ng/mL, such that levels below 50 ng/mL and levels above 150 ng/mL tend to be less effective in treating depressive symptoms (Kaplan and Sadock 1998). Hence, the increase in the patient's nortriptyline level from 91 ng/mL to 218 ng/mL caused by ritonavir's 2D6 inhibition represented a departure from this therapeutic window, leading to the emergence of depressive symptoms. (Incidentally, decreased antidepressant efficacy when nortriptyline blood levels are greater than 150 ng/mL is thought to be unrelated to the presence of any TCA toxicity state.) A reduction in the nortriptyline dosage to 50 mg/day compensated for ritonavir's 2D6 inhibition, leading to a final nortriptyline blood level that was within the therapeutic window and resolution of the patient's depressive symptoms.

References

Kaplan H, Sadock B: Kaplan and Sadock's Synopsis of Psychiatry, Behavioral Sciences/Clinical Psychiatry, 8th Edition. Baltimore, MD, Williams & Wilkins, 1998, p 260

Sawada Y, Ohtani H: Pharmacokinetics and drug interactions of antidepressive agents (in Japanese). Nippon Rinsho 59(8):1539–1545, 2001

von Moltke LL, Greenblatt DJ, Grassi JM, et al: Protease inhibitors as inhibitors of human cytochromes P450: high risk associated with ritonavir. J Clin Pharmacol 38(2):106–111, 1998

"I Get Delirious"

A 45-year-old woman with recurrent major depressive disorder and a family history positive for opioid dependence was being maintained on paroxetine (Paxil), 20 mg/day, with good success. While skiing with her family, she fell and suffered a left humeral fracture. Once hospitalized, she explained to her surgeon that she did not want to receive any conventional narcotics, since she was afraid of becoming "hooked" like her parents. He opted to order tramadol (Ultram), 75 mg po every 4 hours as needed for pain, not to exceed 400 mg/day. She was maintained on her paroxetine as well. She reported only partial relief from her pain, and after 4 days she experienced increasing flushing, diarrhea, muscle twitching, sedation, fevers, and confusion as to time and place. The surgeon suspected an infection and began an appropriate workup. While awaiting blood culture results, the patient experienced a grand mal seizure. She was transferred to the ICU in view of her persistent delirium and vital sign instability, even after the postictal period had passed. Despite her prior directive, she was placed on parenteral morphine, and her tramadol and paroxetine were both discontinued. Her delirium and associated symptoms gradually resolved over the next few days, whereupon she was transferred back to the surgical floor. Her blood culture results were negative, and this prompted the surgeon to consult with the hospital pharmacist. After this consultation, the patient was restarted on her paroxetine, but she was given hydromorphone (Dilaudid) instead of tramadol. She was eventually transitioned to nonsteroidal analgesics without further difficulties.

Discussion

This is an example of a substrate added to an inhibitor.

Tramadol is a substrate of 2D6 (which metabolizes tramadol to a more analgesic compound called M1) (Sindrup et al. 1999), and paroxetine is a strong 2D6 inhibitor (von Moltke et al. 1995). When tramadol was added in standard doses for the treatment of acute pain, the presence of the paroxetine impaired 2D6's ability to efficiently metabolize the transformation of tramadol to M1. This accounted for the patient's report of suboptimal analgesia from an adequate dose of tramadol (Poulsen et al. 1996).

The other contributor to this case was a pharmacodynamic synergy between tramadol's significant serotonergic blockade (and propensity to lower the seizure threshold) (Ultram [package insert] 2000) and that of paroxetine, resulting in a central serotonin syndrome (Lantz et al. 1998) that culminated in a seizure. The discontinuation of the inciting agents led to recovery, and paroxetine was able to be restarted, since it would not impair the analgesic efficacy of hydromorphone.

References

Lantz MS, Buchalter EN, Giambanco V: Serotonin syndrome following the administration of tramadol with paroxetine. Int J Geriatr Psychiatry 13(5):343–345, 1998

Poulsen L, Arendt-Nielsen L, Brosen K, et al: The hypoalgesic effect of tramadol in relation to CYP2D6. Clin Pharmacol Ther 60(6):636–644, 1996

Sindrup SH, Madsen C, Brosen K, et al: The effect of tramadol in painful polyneuropathy in relation to serum drug and metabolite levels. Clin Pharmacol Ther 66(6):636–641, 1999

Ultram (package insert). Raritan, NJ, Ortho-McNeil Pharmaceutical Inc, November 2000

von Moltke LL, Greenblatt DJ, Court MH, et al: Inhibition of alprazolam and desipramine hydroxylation in vitro by paroxetine and fluvoxamine: comparison with other selective serotonin reuptake inhibitor antidepressants. J Clin Psychopharmacol 15(2):125–131, 1995

Chapter 3

3A4 Case Vignettes

The enzyme 3A4 is the "workhorse" of the P450 system. It constitutes roughly 25%–30% of the liver's P450 complement, and there is also a significant 3A4 presence in the gut (DeVane and Nemeroff 2002). There is a vast array of 3A4 substrates, inhibitors, and inducers. Common substrates include tertiary-amine tricyclic antidepressants (TCAs), many antipsychotics, triazolobenzodiazepines, macrolides, calcium-channel blockers, protease inhibitors, most HMG-CoA (3-hydroxy-3-methylglutaryl–coenzyme A) reductase inhibitors, carbamazepine, and steroid compounds. Common inhibitors include some of the selective serotonin reuptake inhibitors (SSRIs) and nefazodone, protease inhibitors, and some of the antifungals, macrolides, and quinolones. Common inducers include several anticonvulsants, rifampin, St. John's wort, and ritonavir (Cozza et al. 2003).

The genetics of 3A4 are complex. Although there are no actual "poor metabolizers," there is vast (10- to 30-fold) variability in the efficiency of 3A4's functioning across the human population (Ketter et al. 1995).

References

Cozza KL, Armstrong SC, Oesterheld JR: Concise Guide to Drug Interaction Principles for Medical Practice: Cytochrome P450s, UGTs, P-Glycoproteins, 2nd Edition. Washington, DC, American Psychiatric Publishing, 2003

DeVane CL, Nemeroff CB: 2002 guide to psychotropic drug interactions. Primary Psychiatry 9(3):28–57, 2002

Ketter TA, Flockhart DA, Post RM, et al: The emerging role of cytochrome P450 3A in psychopharmacology. J Clin Psychopharmacol 15(6):387–398, 1995

Stymied by Statins (I)

A 50-year-old woman had a long-standing history of atypical depression, which was well controlled with fluoxetine (Prozac), 40 mg/day. Routine blood monitoring had unearthed a cholesterol level of 275 mg/dL, and dietary interventions had thus far proven ineffective in correcting this. Her internist prescribed atorvastatin (Lipitor), which was started at 10 mg/day initially and titrated to a dosage of 30 mg/day. However, her cholesterol was still 255 mg/dL, so the dosage of atorvastatin was further raised to 50 mg/day. After 1 month at this dosage, the patient gradually developed extreme fatigue, generalized pruritis, and a mild confusional state, along with a doubling of her liver function tests (LFT)—AST (aspartate transaminase), ALT (alanine transaminase), and GGT (γ-glutamyltransferase)—although these were still within the upper limits of the reported normal range. The atorvastatin was discontinued, and the above symptoms gradually resolved over the next 2 weeks. The fluoxetine was continued throughout. Two months later, her internist tried to add simvastatin (Zocor) to her regimen and titrated this medication to a dosage of 80 mg/day. Within the next 2 weeks, she again experienced extreme fatigue and mild confusion. Blood tests revealed similar elevations of her LFTs, and her creatine phosphokinase (CPK) level was also at the upper end of the normal range. The internist stopped both the simvastatin and the fluoxetine and called his P450-savvy psychiatrist friend, who advised placing the patient on citalopram (Celexa) and pravastatin (Pravachol) after allowing some weeks for these symptoms to abate. The patient tolerated this combination without difficulty, and she enjoyed both antidepressant efficacy and a lowering of her cholesterol level (to 210 mg/dL) (S. Strahan, personal communication, July 2002).

Discussion

This is an example of substrates added to an inhibitor.

Atorvastatin and simvastatin are HMG-CoA reductase inhibitors that are primarily metabolized by 3A4 (Atorvastatin Monograf 2002; Gruer et al. 1999; Neuvonen et al. 1998), whereas pravastatin relies on more heterogeneous hepatic metabolism as well as significant renal excretion (Hatanaka 2000). Fluoxetine is a strong 2D6 inhibitor, and its active metabolite, norfluoxetine, is a moderate 3A4 inhibitor (Greenblatt et al. 1999; Stevens and Wrighton 1993; von Moltke et al. 1995). Thus, when atorvastatin and simvastatin were added to the fluoxetine, their metabolism at 3A4 was inhibited and their resultant blood levels were greater than would have been expected in this patient at the doses prescribed. Thus, symptoms of HMG-CoA reductase inhibitor toxicity emerged (elevated LFTs, mild delirium, mild rhabdomyolysis). However, when pravastatin (whose metabolism is difficult to inhibit) was combined with citalopram (which lacks clinically significant P450 inhibitory capabilities) (Brosen and Naranjo 2001), a tolerable and effective regimen was achieved.

References

Atorvastatin Monograf Ubat 180, pp 1–13, March 2002. Available at: http://www.bpfk.org/html/(180)%20Atorvastatin%20-%20March02.pdf. Accessed September 7, 2002

Brosen K, Naranjo CA: Review of pharmacokinetic and pharmacodynamic interaction studies with citalopram. Eur Neuropsychopharmacol 11(4):275–283, 2001

Greenblatt DJ, von Moltke LL, Harmatz JS, et al: Human cytochromes and some newer antidepressants: kinetics, metabolism, and drug interactions. J Clin Psychopharmacol 19(5, suppl 1):23S–35S, 1999

Gruer PJ, Vega JM, Mercuri MF, et al: Concomitant use of cytochrome P450 3A4 inhibitors and simvastatin. Am J Cardiol 84(7):811–815, 1999

Hatanaka T: Clinical pharmacokinetics of pravastatin: mechanisms of pharmacokinetic events. Clin Pharmacokinet 39(6):397–412, 2000

Neuvonen PJ, Kantola T, Kivisto KT: Simvastatin but not pravastatin is very susceptible to interaction with the CYP3A4 inhibitor itraconazole. Clin Pharmacol Ther 63(3):332–341, 1998

Stevens JC, Wrighton SA: Interaction of the enantiomers of fluoxetine and norfluoxetine with human liver cytochromes P450. J Pharmacol Exp Ther 266(2): 964–971, 1993

von Moltke LL, Greenblatt DJ, Court MH, et al: Inhibition of alprazolam and desipramine hydroxylation in vitro by paroxetine and fluvoxamine: comparison with other selective serotonin reuptake inhibitor antidepressants. J Clin Psychopharmacol 15(2):125–131, 1995

A Fatal Case of Bronchitis

A 55-year-old woman had been prescribed pimozide (Orap), 4 mg/day, by her dermatologist for treatment of delusional parasitosis. The patient had responded well to this dose, which she tolerated with only mild complaints of stiffness. One winter, she caught the flu and began to cough frequently and painfully. When she developed a fever of 101.3°, she visited her internist, who prescribed clarithromycin (Biaxin), 500 mg bid for 10 days. On day 5, the patient went to sleep and never woke up. The postmortem examination led to a diagnosis of a fatal arrhythmia (K.L. Cozza, personal communication, May 2002).

Discussion

This is an example of an inhibitor added to a substrate.

Pimozide is a 3A4 substrate, and clarithromycin is a strong 3A4 inhibitor (Desta et al. 1999). The addition of the clarithromycin significantly impaired the ability of 3A4 to efficiently metabolize the pimozide. This caused the blood level of pimozide to increase significantly, even though the dosage of pimozide had not been changed. Pimozide can increase the QTc interval in a dose-dependent manner. Thus, when the pimozide blood level increased, the QTc interval likely increased significantly as well. Clarithromycin can also increase the QTc interval. Hence, there was a pharmacodynamic interaction (additive QTc interval prolongation) in addition to the pharmacokinetic one (3A4 inhibition). The result was that the patient likely experienced a large QTc prolongation, predisposing to a torsades de pointes arrhythmia, leading to ventricular fibrillation and death.

References

Desta Z, Kerbusch T, Flockhart DA: Effect of clarithromycin on the pharmacokinetics and pharmacodynamics of pimozide in healthy poor and extensive metabolizers of cytochrome P450 2D6 (CYP2D6). Clin Pharmacol Ther 65(1): 10–20, 1999

Natural Disaster (I)

A 35-year-old man had just undergone a successful cardiac transplantation procedure to treat his debilitating cardiomyopathy. He was started on cyclosporine (Sandimmune) so that his immune system would not reject his new heart. He began to experience some depressive symptoms (moderate dysphoria, insomnia, poor appetite, poor energy) in the weeks following this procedure. He decided to obtain his own supply of St. John's wort *(Hypericum perforatum)* and began taking it at the dosage indicated by the pharmacist. Since he was taking so many "artificial" drugs, he wanted to take a "natural" remedy for depression, believing that this would be a "healthier" alternative to the usual antidepressants. Roughly

1 month later, he began to experience increasing fatigue, dyspnea, and orthopnea, as well as intermittent fevers. When he reported these symptoms to his cardiologist, he was admitted to the hospital, where he was found to be in congestive heart failure with an ejection fraction of 20%. His cyclosporine blood level was far below the minimum effective value (Ruschitzka et al. 2000).

Discussion

This is an example of an inducer added to a substrate.

Cyclosporine is a 3A4 substrate (Kronbach et al. 1988), and St. John's wort is a 3A4 inducer (Moore et al. 2000; Roby et al. 2000). When the St. John's wort was added to the cyclosporine, there was an increase in the amount of 3A4 available to metabolize the cyclosporine. Over the next several weeks, this led to more efficient metabolism of the cyclosporine and a resultant decrease in the blood level of cyclosporine from its previous baseline level. Since the cyclosporine was inhibiting the patient's immune response to the foreign donor heart, this decrease in the cyclosporine blood level led to a reactivation of his immune response and the resulting organ rejection.

There is also a likely contribution to this decrease in cyclosporine blood level from alterations in the functioning of the P-glycoprotein transporter. Cyclosporine is a P-glycoprotein substrate (Yokogawa et al. 2002), and St. John's wort is a P-glycoprotein inducer (Hennessy et al. 2002). Thus, the St. John's wort increased the number of P-glycoprotein transporters and hence increased the overall activity of P-glycoprotein, which led to more cyclosporine being extruded from enterocytes back into the gut lumen, where it was excreted rather than absorbed. This increase in excretion and decrease in absorption of cyclosporine also contributed to the decrease in the cyclosporine blood level following the addition of St. John's wort.

References

Hennessy M, Kelleher D, Spiers JP, et al: St Johns wort increases expression of P-glycoprotein: implications for drug interactions. Br J Clin Pharmacol 53(1): 75–82, 2002

Kronbach T, Fischer V, Meyer UA: Cyclosporine metabolism in human liver: identification of a cytochrome P-450III gene family as the major cyclosporine-metabolizing enzyme explains interactions of cyclosporine with other drugs. Clin Pharmacol Ther 43(6):630–635, 1988

Moore LB, Goodwin B, Jones SA, et al: St. John's wort induces hepatic drug metabolism through activation of the pregnane X receptor. Proc Natl Acad Sci USA 97(13):7500–7502, 2000
Roby CA, Anderson GD, Kantor E, et al: St. John's wort: effect on CYP3A4 activity. Clin Pharmacol Ther 67(5):451–457, 2000
Ruschitzka F, Meier PJ, Turina M, et al: Acute heart transplant rejection due to Saint John's wort (letter). Lancet 355(9203):548–549, 2000
Yokogawa K, Shimada T, Higashi Y, et al: Modulation of *mdr1a* and *CYP3A* gene expression in the intestine and liver as possible cause of changes in the cyclosporin A disposition kinetics by dexamethasone. Biochem Pharmacol 63(4): 777–783, 2002

Surprising Sedation

A 22-year-old woman has carried the diagnoses of both bipolar I disorder and panic disorder. Her symptoms have been well controlled on a regimen of carbamazepine (Tegretol), 800 mg/day, and alprazolam (Xanax), 4 mg/day. One of her routine complete blood counts revealed an evolving leukopenia. After discussing risks and benefits, she and her psychiatrist decided to quickly discontinue her carbamazepine and begin a trial of lithium. Initial titration of the lithium to a dosage of 900 mg/day took place without incident, and this dosage was well tolerated. However, within 3 weeks of discontinuing the carbamazepine, she experienced marked sedation, which interfered with her ability to do her job. Her psychiatrist decreased her alprazolam to 1.5 mg/day, which continued to effectively prevent panic attacks and which she tolerated without sedation (Pies 2002).

Discussion

This is an example of reversal of induction.

Alprazolam is a 3A4 substrate, as are the other triazolobenzodiazepines (triazolam, midazolam, and estazolam) (Dresser et al. 2000). Carbamazepine is a 3A4 inducer (Spina et al. 1996). When carbamazepine and alprazolam were coadministered, the alprazolam was titrated to stable clinical effect (control of panic attacks without sedation) at a dosage that took into account a stable degree of 3A4 induction by carbamazepine. With the discontinuation of the carbamazepine, there was a reversal of 3A4 induction. The resulting decrease in available 3A4 led to less active metabolism of the alprazolam, yielding higher alprazolam blood levels and subsequent oversedation.

References

Dresser GK, Spence JD, Bailey DG: Pharmacokinetic-pharmacodynamic consequences and clinical relevance of cytochrome P450 3A4 inhibition. Clin Pharmacokinet 38(1):41–57, 2000

Pies R: Cytochromes and beyond: drug interactions in psychiatry. Psychiatric Times, May 2002, pp 48–51

Spina E, Pisani F, Perucca E: Clinically significant pharmacokinetic drug interactions with carbamazepine: an update. Clin Pharmacokinet 31(3):198–214, 1996

Two Rights Can Make a Wrong

A 42-year-old woman with anxious depression achieved good symptom control on a regimen of citalopram (Celexa), 40 mg/day, and buspirone (BuSpar), 30 mg/day. However, she described to her psychiatrist that she had been experiencing a decrease in libido since starting the citalopram. After thorough discussion, she and her psychiatrist decided to add nefazodone and taper the citalopram in a crossover titration strategy. The rationale was that nefazodone would be helpful both as an agent to treat anxious depression and as an antidepressant that would be much less likely to impair libido. Within 2 weeks, the patient reached a nefazodone dose of 200 mg bid. However, even before this dose was reached, the patient reported new-onset dizziness, nausea, and severe fatigue and sedation. The psychiatrist consulted her hospital pharmacist and then abruptly discontinued the buspirone. She also opted to hold the nefazodone for 1 week. After that week, the patient's symptoms resolved, and the nefazodone was again titrated to an eventual dosage of 500 mg/day. This dosage was well tolerated and eventually proved effective.

Discussion

This is an example of an inhibitor added to a substrate.

Buspirone is a 3A4 substrate (Kivisto et al. 1997), and nefazodone is a strong 3A4 inhibitor (von Moltke et al. 1996). Thus, when the nefazodone was added, the metabolism of buspirone by 3A4 was markedly impaired, with the result that the blood level of buspirone increased significantly. Published data indicate that the addition of nefazodone to buspirone, at routine dosages of each, can increase buspirone blood levels up to 20-fold! (See Serzone [package insert] 2001 for details.) In this case, the increase in the buspirone blood level yielded the side effects of nausea, dizziness, and sedation.

References

Kivisto KT, Lamberg TS, Kantola T, et al: Plasma buspirone concentrations are greatly increased by erythromycin and itraconazole. Clin Pharmacol Ther 62(3):348–354, 1997

Serzone (package insert). Princeton, NJ, Bristol-Myers Squibb, February 2001

von Moltke LL, Greenblatt DJ, Harmatz JS, et al: Triazolam biotransformation by human liver microsomes in vitro: effects of metabolic inhibitors and clinical confirmation of a predicted interaction with ketoconazole. J Pharmacol Exp Ther 276(2):370–379, 1996

Horizontal

An 79-year-old man with long-standing anxious depression was undergoing a trial of mirtazapine (Remeron), 30 mg/day. He was also taking nifedipine (Procardia) for his hypertension. The patient was not able to tolerate the mirtazapine because of sedation. His psychiatrist decided to discontinue the mirtazapine and begin a trial of fluoxetine (Prozac). Roughly 3 weeks after reaching a fluoxetine dosage of 20 mg/day, the patient attempted to get out of bed, whereupon he had a syncopal episode. His son heard him fall and found him lying on his back on the floor. The son called an ambulance, and when the paramedics took the man's supine blood pressure, it was 80/50 mm Hg (Azaz-Livshits and Danenberg 1997).

Discussion

This is an example of an inhibitor added to a substrate.

Nifedipine is a 3A4 substrate (Iribarne et al. 1997), and fluoxetine produces moderate inhibition of 3A4 through its active metabolite, norfluoxetine (Greenblatt et al. 1999; von Moltke et al. 1995). At the patient's baseline, the blood level of nifedipine appropriately controlled his hypertension without producing orthostasis. However, the introduction of fluoxetine led to a significant decrease in the ability of 3A4 to efficiently metabolize nifedipine. This resulted in an increase in the blood level of nifedipine, even though the dosage of nifedipine had not been changed, leading to the orthostasis-induced syncopal episode described above.

References

Azaz-Livshits TL, Danenberg HD: Tachycardia, orthostatic hypotension and profound weakness due to concomitant use of fluoxetine and nifedipine. Pharmacopsychiatry 30(6):274–275, 1997

Greenblatt DJ, von Moltke LL, Harmatz JS, et al: Human cytochromes and some newer antidepressants: kinetics, metabolism, and drug interactions. J Clin Psychopharmacol 19 (5, suppl 1):23S–35S, 1999

Iribarne C, Dreano Y, Bardou LG, et al: Interaction of methadone with substrates of human hepatic cytochrome P450 3A4. Toxicology 117(1):13–23, 1997

von Moltke LL, Greenblatt DJ, Court MH, et al: Inhibition of alprazolam and desipramine hydroxylation in vitro by paroxetine and fluvoxamine: comparison with other selective serotonin reuptake inhibitor antidepressants. J Clin Psychopharmacol 15(2):125–131, 1995

Sedation, Terminable and Interminable

A 40-year-old man was referred by his internist to a gastroenterologist for his first-ever screening colonoscopy. His only medications were diltiazem (Cardizem) for hypertension and a multivitamin. The gastroenterologist typically used midazolam (Versed) as his sedative agent during this procedure. The colonoscopy proceeded uneventfully, but following the procedure, the patient remained unconscious almost 3 hours longer than the average patient. Since this was the final "scope" of the day, this obligated the gastroenterologist and a nurse to wait with the patient until he finally woke up. Even then, the patient needed his spouse to come and drive him home because of residual sedation.

Discussion

This is an example of a substrate added to an inhibitor.

Midazolam is a 3A4 substrate (Dresser et al. 2000; von Moltke et al. 1996), and diltiazem is a 3A4 inhibitor (Sutton et al. 1997). Midazolam was administered in a standard dose that did not take into account the presence of a significant 3A4 inhibitor. Since the diltiazem was present, the ability of 3A4 to efficiently metabolize the midazolam was significantly impaired. This resulted in a significantly higher than expected blood level of midazolam, which yielded excessive sedation and much inconvenience for all concerned. In some surgical procedures in which intubation is required, this interaction has resulted in significant delays (2–3 hours) before patients could be safely extubated (Ahonen et al. 1996).

References

Ahonen J, Olkkola KT, Salmenpera M, et al: Effect of diltiazem on midazolam and alfentanil disposition in patients undergoing coronary artery bypass grafting. Anesthesiology 85(6):1246–1252, 1996

Dresser GK, Spence JD, Bailey DG: Pharmacokinetic-pharmacodynamic consequences and clinical relevance of cytochrome P450 3A4 inhibition. Clin Pharmacokinet 38(1):41–57, 2000

Sutton D, Butler AM, Nadin L, et al: Role of CYP3A4 in human hepatic diltiazem N-demethylation: inhibition of CYP3A4 activity by oxidized diltiazem metabolites. J Pharmacol Exp Ther 282(1):294–300, 1997

von Moltke LL, Greenblatt DJ, Schmider J, et al: Midazolam hydroxylation by human liver microsomes in vitro: inhibition by fluoxetine, norfluoxetine, and by azole antifungal agents. J Clin Pharmacol 36(9):783–791, 1996

Tuberculous Anxiety

A 35-year-old man diagnosed with generalized anxiety disorder (GAD) was receiving good benefit from buspirone (BuSpar), 60 mg/day. After traveling abroad for several months, he began to experience a frequent and painful dry cough, as well as malaise and intermittent fever. He visited his internist, and a workup eventually revealed active tuberculosis. Among other medications, he was placed on rifampin (in the form of Rifadin) at standard doses. Within the next 2 weeks, he experienced greatly heightened generalized anxiety. Initially interpreting the patient's increased anxiety as being related to his recently finding out about the tuberculosis, the internist referred the patient to a psychologist for supportive counseling. When this did not prove helpful, the internist reassessed the patient's medication regimen and consulted his hospital pharmacist. The answer he received prompted him to add clonazepam (Klonopin), which eventually provided significant relief for the patient's anxiety when titrated to a dose of 1.5 mg bid.

Discussion

This is an example of an inducer added to a substrate.

Buspirone is a 3A4 substrate (Kivisto et al. 1997). Rifampin is a "pan-inducer," which means that it induces virtually all of the major P450 enzymes, including 3A4 (Strayhorn et al. 1997). The addition of rifampin led to an increased amount of 3A4, which led to more efficient metabolism of the buspirone. This resulted in a greatly decreased blood level of buspirone within 1–2 weeks of starting the rifampin, which led to a relapse of the patient's active GAD symptoms. Studies have demonstrated that this specific interaction may result in maximum buspirone blood levels that are only 9% (!) of baseline values, and an AUC (area under the curve) that is only 15% of baseline (Kivisto et al. 1999). When the pharmacist shared this information with the internist, the internist wisely decided that treating the patient's anxiety by increasing the dosage of buspirone would have been impractical, potentially requiring several hundred milligrams per day, and he instead chose to add clonazepam, which provided symptomatic relief.

References

Kivisto KT, Lamberg TS, Kantola T, et al: Plasma buspirone concentrations are greatly increased by erythromycin and itraconazole. Clin Pharmacol Ther 62(3):348–354, 1997

Kivisto KT, Lamberg TS, Neuvonen PJ: Interactions of buspirone with itraconazole and rifampicin: effects on the pharmacokinetics of the active 1-(2-pyrimidinyl)-piperazine metabolite of buspirone. Pharmacol Toxicol 84(2):94–97, 1999

Strayhorn VA, Baciewicz AM, Self TH: Update on rifampin drug interactions, III. Arch Intern Med 157(21):2453–2458, 1997

Sleepy

A 21-year-old college student was taking alprazolam (Xanax), 1 mg tid, for panic disorder, which provided good relief and which she tolerated without difficulty. Numerous separation issues and impending life changes, as well as final examination pressures, led to an especially stressful 6 weeks before her planned graduation. She began to experience increasing dysphoria and anhedonia; poor sleep, appetite, and concentration; and heightened generalized anxiety—which led the patient and her psychiatrist to consider adding an antidepressant to her regimen. They decided to try nefazodone (Serzone). Within 2 days of reaching a dose of only 100 mg bid, the patient began to experience marked sedation, which was interfering with her ability to study. The psychiatrist discontinued the alprazolam, which led to a resolution of the sedation, but after 3 days, there was a reemergence of her panic attacks. Her alprazolam was then restarted at half the previous dose (0.5 mg tid), which proved effective for the panic attacks without producing excessive sedation.

Discussion

This is an example of an inhibitor added to a substrate.

Alprazolam is a 3A4 substrate (Dresser et al. 2000), and nefazodone is a strong 3A4 inhibitor (von Moltke et al. 1996). The addition of nefazodone led to a significant reduction in the ability of 3A4 to efficiently metabolize alprazolam. This resulted in a significant increase (roughly doubling) of the blood level of alprazolam, even though the alprazolam dose had initially remained constant (DeVane and Nemeroff 2002). A halving of the alprazolam dosage compensated for this inhibition, leading to therapeutic efficacy without oversedation. Since both of these agents are independently sedating, additive receptor effects (pharmacodynamic effects) may have also contributed to this interaction.

References

DeVane CL, Nemeroff CB: 2002 guide to psychotropic drug interactions. Primary Psychiatry 9(3):28–57, 2002

Dresser GK, Spence JD, Bailey DG: Pharmacokinetic-pharmacodynamic consequences and clinical relevance of cytochrome P450 3A4 inhibition. Clin Pharmacokinet 38(1):41–57, 2000

von Moltke LL, Greenblatt DJ, Harmatz JS, et al: Triazolam biotransformation by human liver microsomes in vitro: effects of metabolic inhibitors and clinical confirmation of a predicted interaction with ketoconazole. J Pharmacol Exp Ther 276(2):370–379, 1996

Preoccupied

A 31-year-old man with a long history of anxious depression and hypochondriasis had maintained a good response over the past 3 years while taking nefazodone (Serzone), 600 mg/day (tried first) and alprazolam (Xanax), 1 mg bid (added later). However, after hearing on the news that nefazodone can cause “liver damage,” he presented to his psychiatrist in a highly anxious state. The patient and his psychiatrist discussed the treatment issues, and the patient had a hepatic function panel drawn, which was entirely unremarkable. Despite this reassurance, the patient could not stop ruminating over the concern that he was slowly “killing” his liver. After much discussion, the patient and his psychiatrist agreed to add citalopram (Celexa), titrating to a dosage of 20 mg/day and then gradually tapering the nefazodone at a rate of 100 mg/week. This cross-titration strategy seemed to be working well, until the patient reached a nefazodone dosage of 200 mg/day, whereupon he began to experience increasing anxiety. The anxiety did not abate despite reassurance. Alprazolam was increased to 1.5 mg bid, which provided only moderate relief. Within 3 days of the dosage of nefazodone being further decreased to 100 mg/day, the patient became intensely and pervasively anxious. He was also sweating and tremulous, and he measured his pulse rate at 100 beats per minute. His distress was so great that he agreed to briefly increase the dosage of nefazodone back to 400 mg/day in the hope of reversing whatever process had led to such a drastic increase in his anxiety level. Once a nefazodone dosage of 400 mg/day was achieved, his anxiety did indeed abate. His psychiatrist eventually consulted a pharmacist, who revealed what the difficulty had been. The patient was eventually able to tolerate a nefazodone taper as long as it was accompanied by steady increases in his alprazolam dosage.

Discussion

This is an example of reversal of inhibition.

As in the previous example (see "Sleepy," p. 45), alprazolam is a 3A4 substrate (Dresser et al 2000), and nefazodone is a strong 3A4 inhibitor (von Moltke et al. 1996). Initially, the nefazodone and alprazolam were being coadministered at dosages that provided treatment efficacy without generating side effects. However, this effective dosage of alprazolam relied on nefazodone's inhibiting the ability of 3A4 to efficiently metabolize the alprazolam. With the tapering of the nefazodone, 3A4 was able to more efficiently metabolize the alprazolam, which led to a decrease in the blood level of alprazolam. This led to a mild benzodiazepine withdrawal state, resulting in heightened anxiety, tremor, diaphoresis, and tachycardia. A greater increase in the dosage of alprazolam (to a total of 4 mg/day or more) would likely have prevented these withdrawal symptoms. Replacing the nefazodone again introduced enough 3A4 inhibition to produce an alprazolam blood level that was sufficient to correct this withdrawal state.

References

Dresser GK, Spence JD, Bailey DG: Pharmacokinetic-pharmacodynamic consequences and clinical relevance of cytochrome P450 3A4 inhibition. Clin Pharmacokinet 38(1):41–57, 2000

von Moltke LL, Greenblatt DJ, Harmatz JS, et al: Triazolam biotransformation by human liver microsomes in vitro: effects of metabolic inhibitors and clinical confirmation of a predicted interaction with ketoconazole. J Pharmacol Exp Ther 276(2):370–379, 1996

Ataxic

A 32-year-old man with a history of bipolar I disorder had been stably maintained on carbamazepine (Tegretol), 400 mg bid, with blood levels ranging between 8 and 10 μg/mL. One winter, he developed an atypical pneumonia, for which his internist prescribed erythromycin (E-mycin), 500 mg qid for 10 days. After 4 days, the patient was experiencing increased sedation, ataxia, slurred speech, and jerking movements of his arms, which prompted a visit to the local ER. While there, his carbamazepine blood level was drawn and was revealed to be 25.6 μg/mL. A subsequent trough level was 19.2 μg/mL. An electrocardiogram (ECG) was unremarkable. The ER physician discontinued the erythromycin and had the patient begin taking azithromycin (Zithromax). He also instructed the patient to skip his next two doses of carbamazepine and then resume his normal regimen.

Discussion

This is an example of an inhibitor added to a substrate.

Carbamazepine is primarily a 3A4 substrate, with 1A2 and 2C9 making minor contributions to carbamazepine's metabolism (Spina et al. 1996). Erythromycin is a 3A4 inhibitor (Pai et al. 2000). The addition of the erythromycin led to a significant impairment in the ability of 3A4 to effectively contribute to the metabolism of carbamazepine. Since the activity of 1A2 and 2C9 was not sufficient to compensate for this effect, the inhibition of 3A4 resulted in roughly a doubling of the carbamazepine blood level, even though the carbamazepine dosage had not been changed (Spina et al. 1996). This caused the emergence of typical signs of carbamazepine toxicity (sedation, ataxia, slurred speech, jerking arms).

Although it did not become a significant issue here, this case also contained an example of a substrate added to an inducer. Erythromycin is a substrate of 3A4 (Erythromycin [package insert] 2000) in addition to being an inhibitor, while carbamazepine is an inducer of 3A4 (Spina et al. 1996) in addition to being a substrate. The erythromycin was being dosed in a standard manner that did not take into account the presence of a 3A4 inducer (carbamazepine). The presence of such an inducer necessitates increased dosing of substrates, including erythromycin, if one wishes to generate blood levels comparable to those that a normal dosing regimen would generate in the absence of such an inducer. Therefore, if this antibiotic treatment course had reached its 10-day end point, it might well have proven to be an ineffective dosage of erythromycin, and the pneumonia might not have been adequately treated.

There is also a possible P-glycoprotein contribution to this interaction. Carbamazepine is a P-glycoprotein substrate (Potschka et al. 2001), and erythromycin is a P-glycoprotein inhibitor (Kiso et al. 2000). Thus, the addition of the erythromycin caused a decrease in the activity of P-glycoprotein in extruding carbamazepine from enterocytes back into the gut lumen, where it would have been excreted rather than absorbed. This led to increased absorption of carbamazepine and, thus, an increased blood level of carbamazepine.

References

Erythromycin (package insert). North Chicago, IL, Abbott Laboratories, 2000

Kiso S, Cai SH, Kitaichi K, et al: Inhibitory effect of erythromycin on P-glycoprotein–mediated biliary excretion of doxorubicin in rats. Anticancer Res 20(5A): 2827–2834, 2000

Pai MP, Graci DM, Amsden GW: Macrolide drug interactions: an update. Ann Pharmacother 34(4):495–513, 2000

Potschka H, Fedrowitz M, Loscher W: P-glycoprotein and multidrug resistance–associated protein are involved in the regulation of extracellular levels of the major antiepileptic drug carbamazepine in the brain. Neuroreport 12(16): 3557–3560, 2001

Spina E, Pisani F, Perucca E: Clinically significant pharmacokinetic drug interactions with carbamazepine: an update. Clin Pharmacokinet 31(3):198–214, 1996

Paranoia

A 28-year-old construction worker who carried the diagnosis of schizophrenia had done quite well for the past 2 years on a regimen of quetiapine (Seroquel), 600 mg/day. One day, he fell from a scaffold and suffered significant closed head injury. During the recovery period, he experienced several tonic-clonic seizures. These seizures prompted the consulting neurologist to have him start taking phenytoin (Dilantin), 300 mg/day. His seizure activity was rapidly brought under control with the phenytoin. He was then transitioned to a rehabilitation facility. However, in subsequent weeks, he began to grow more flagrantly paranoid. He became fearful of sleeping because of the belief that the staff would "experiment" on him while he slept. He also refused to eat or drink anything that he had not opened himself. The quetiapine dosage was increased to 800 mg/day, but this did not halt the progression of his paranoid delusions. Finally, after barricading himself in his room and becoming acutely threatening, he was transferred to an inpatient psychiatric facility. The P450-savvy inpatient psychiatrist switched the patient from quetiapine to ziprasidone (Geodon), titrating to a dosage of 160 mg/day. The patient tolerated the medication at this dosage without difficulty, and his delusions remitted within another 3 weeks.

Discussion

This is an example of an inducer added to a substrate.

Quetiapine is primarily a 3A4 substrate, with a minor contribution from 2D6 (DeVane and Nemeroff 2001), and phenytoin is a strong inducer of several P450 enzymes, including 3A4 (Anderson 1998). The addition of the phenytoin led to an increase in the amount of 3A4 available to metabolize the quetiapine, resulting in significantly more rapid metabolism of quetiapine. This led to a sharp decrease in the blood level of quetiapine. Studies indicate that this interaction can increase the clearance of quetiapine fivefold (Wong et al. 2001). Hence, the increase in dosage of quetiapine from 600 mg/day to 800 mg/day was not able to compensate for this effect. The significant decrease in quetiapine blood level that ensued, over the course of several weeks, led to the emergence of paranoid delusions and accompanying problematic behaviors.

References

Anderson GD: A mechanistic approach to antiepileptic drug interactions. Ann Pharmacother 32(5):554–563, 1998

DeVane CL, Nemeroff CB: Clinical pharmacokinetics of quetiapine: an atypical antipsychotic. Clin Pharmacokinet 40(7):509–522, 2001

Wong YW, Yeh C, Thyrum PT: The effects of concomitant phenytoin administration on the steady-state pharmacokinetics of quetiapine. J Clin Psychopharmacol 21(1):89–93, 2001

Crash

A 38-year-old woman with chronic panic disorder had responded well to a regimen of venlafaxine (Effexor), 150 mg/day, and alprazolam, 1 mg tid. One winter she developed bronchitis. Her painful coughing and fever prompted her to visit her internist, who prescribed clarithromycin, 500 mg bid for 10 days. Over the next several days, she experienced increasing fatigue and sedation. On the night of day 4 of her clarithromycin regimen, she was prevented from sleeping by her newborn child, who also had an upper respiratory infection. She then drove to her job and worked until

7 P.M., having taken her final alprazolam dose of the day at 6 P.M. During the ride home, she fell asleep at the wheel and crashed her car into a tree.

Discussion

This is an example of an inhibitor added to a substrate.

Alprazolam is a 3A4 substrate (Dresser et al. 2000), and clarithromycin is a strong 3A4 inhibitor (Desta et al. 1999). At baseline, the patient's alprazolam dose of 1 mg tid provided relief from panic attacks without producing excessive sedation. However, with the addition of the clarithromycin, the ability of 3A4 to efficiently metabolize the alprazolam was significantly impaired, resulting in a significantly increased blood level of alprazolam, even though the dose had not been changed (DeVane and Nemeroff 2002). This led to her increasing daytime sedation, culminating in her car crash.

References

Desta Z, Kerbusch T, Flockhart DA: Effect of clarithromycin on the pharmacokinetics and pharmacodynamics of pimozide in healthy poor and extensive metabolizers of cytochrome P450 2D6 (CYP2D6). Clin Pharmacol Ther 65(1): 10–20, 1999

DeVane CL, Nemeroff CB: 2002 guide to psychotropic drug interactions. Primary Psychiatry 9(3):28–57, 2002

Dresser GK, Spence JD, Bailey DG: Pharmacokinetic-pharmacodynamic consequences and clinical relevance of cytochrome P450 3A4 inhibition. Clin Pharmacokinet 38(1):41–57, 2000

Bitter Fruit

A 26-year-old man with bipolar I disorder had maintained acceptable control on a regimen of lithium (Eskalith), 1,200 mg/day (most recent level = 0.8 mEq/L), and carbamazepine (Tegretol), 400 mg bid (most recent level = 11 μg/mL). In his ongoing efforts to adopt a "healthier" lifestyle, he decided to stop drinking sodas and he began to consume more than 16 ounces per day of grapefruit juice. Over the next week, he became increasingly sedated, mildly tremulous, and significantly nauseated. After vomiting twice in 2 days, he complained to his psychiatrist. The psychiatrist ordered a carbamazepine level, which was found to be 17.5 μg/mL. After discussing the patient's recent dietary changes, the psychiatrist suggested skipping the next two doses of carbamazepine and switching from grapefruit juice to water.

Discussion

This is an example of an inhibitor added to a substrate.

Carbamazepine is primarily a 3A4 substrate, with 1A2 and 2C9 making minor contributions to carbamazepine's metabolism (Spina et al. 1996). Grapefruit juice is an inhibitor of hepatic 1A2 (Fuhr 1998; Fuhr et al. 1993) and a moderate inhibitor of intestinal, but not hepatic, 3A4 as well (Fuhr 1998; He et al. 1998). The addition of grapefruit juice to the patient's diet impaired the ability of 1A2 and, more importantly, intestinal 3A4 to contribute to the metabolism of carbamazepine, thus raising the blood level of carbamazepine even though the dosage had not been changed (Garg et al. 1998). This increase in the carbamazepine blood level led to the patient's new nausea and vomiting, tremor, and sedation. This carbamazepine-grapefruit juice interaction is likely mediated both through P450 enzyme inhibition (3A4 [mostly] and 1A2) and through inhibition of the P-glycoprotein transport system. As it happens, carbamazepine is a substrate of the P-glycoprotein transporter (Potschka et al. 2001), and grapefruit is an inhibitor of this transporter (Wang et al. 2001). Thus, in addition to 3A4 and 1A2 inhibition, grapefruit juice also raises carbamazepine levels by impairing the ability of P-glycoprotein to extrude carbamazepine from enterocytes back into the gut lumen, where it would then be excreted rather than absorbed.

Even though the interaction in this case was more bothersome than dangerous, there is evidence that grapefruit juice can significantly elevate the blood levels of calcium-channel blockers and cyclosporine (Fuhr 1998), possibly with dire consequences. It is also prudent to avoid combining grapefruit juice with clozapine or with tertiary-amine TCAs. Since grapefruit juice is obviously not a regulated substance, it is important to ask specifically about patient consumption of grapefruit or grapefruit juice, just as one would inquire about any other agent on a medication list. Concerns regarding unregulated grapefruit juice consumption have led some astute clinicians to have it removed from hospital menus.

References

Fuhr U: Drug interactions with grapefruit juice: extent, probable mechanism and clinical relevance. Drug Saf 18(4):251–272, 1998

Fuhr U, Klittich K, Staib AH, et al: Inhibitory effect of grapefruit juice and its bitter principal, naringenin, on CYP1A2 dependent metabolism of caffeine in man. Br J Clin Pharmacol 35(4):431–436, 1993

Garg SK, Kumar N, Bhargava VK, et al: Effect of grapefruit juice on carbamazepine bioavailability in patients with epilepsy. Clin Pharmacol Ther 64(3): 286–288, 1998

He K, Iyer KR, Hayes RN, et al: Inactivation of cytochrome P450 3A4 by bergamottin, a component of grapefruit juice. Chem Res Toxicol 11(4):252–259, 1998

Potschka H, Fedrowitz M, Loscher W: P-glycoprotein and multidrug resistance–associated protein are involved in the regulation of extracellular levels of the major antiepileptic drug carbamazepine in the brain. Neuroreport 12(16): 3557–3560, 2001

Spina E, Pisani F, Perucca E: Clinically significant pharmacokinetic drug interactions with carbamazepine: an update. Clin Pharmacokinet 31(3):198–214, 1996

Wang EJ, Casciano CN, Clement RP, et al: Inhibition of P-glycoprotein transport function by grapefruit juice psoralen. Pharm Res 18(4):432–438, 2001

Unintended Fertility

A 20-year-old college student was using an oral contraceptive containing ethinylestradiol as her means of preventing pregnancy. She had a boyfriend with whom she had engaged in intercourse approximately twice a week for the past 10 months. The oral contraceptive was their only method of birth control. During a period of increased stress and dysphoria, she purchased some St. John's wort (*Hypericum perforatum*) at the local drugstore and began taking it per the attached instructions. Three months later, she discovered that she was pregnant.

Discussion

This is an example of an inducer added to a substrate.

Ethinylestradiol is a 3A4 substrate (Guengerich 1990), and St. John's wort is a 3A4 inducer (Moore et al. 2000; Roby et al. 2000). In the weeks following the addition of the St. John's wort, more 3A4 was produced and therefore available to metabolize the ethinylestradiol. This led to lower blood levels of ethinylestradiol, with a subsequent loss of efficacy as a reliable method of birth control (J.R. Oesterheld, personal communication, May 2002). In the face of regular sexual activity, this resulted in an unintended pregnancy.

References

Guengerich FP: Metabolism of 17α-ethynylestradiol in humans. Life Sci 47(22): 1981–1988, 1990

Moore LB, Goodwin B, Jones SA, et al: St. John's wort induces hepatic drug metabolism through activation of the pregnane X receptor. Proc Natl Acad Sci USA 97(13):7500–7502, 2000

Roby CA, Anderson GD, Kantor E, et al: St. John's wort: effect on CYP3A4 activity. Clin Pharmacol Ther 67(5):451–457, 2000

Gradual Withdrawal (I)

A 25-year-old man with a history of opioid dependence had been enrolled in a methadone maintenance program for the past 2 years. His regular dosage of methadone (Dolophine) was 50 mg/day. While taking this regimen, he developed trigeminal neuralgia, and a consulting neurologist prescribed carbamazepine (Tegretol), 600 mg/day, to which the neuralgia promptly responded. However, over the next several weeks, the patient developed typical symptoms of opioid withdrawal (muscle cramping, diarrhea, piloerection, nausea, and insomnia). The clinic psychiatrist was consulted. After reviewing the patient's history, the psychiatrist increased the dosage of the patient's methadone to 80 mg/day and prevailed on his neurologist colleague to change the carbamazepine to gabapentin (Neurontin). The patient's withdrawal was ameliorated by the increase in methadone dosage, and over the next 3 weeks the methadone dosage was tapered back to 50 mg/day, with no further adverse effects.

Discussion

This is an example of an inducer added to a substrate.

Methadone is a 3A4 substrate (Iribarne et al. 1996), and carbamazepine is a 3A4 inducer (Spina et al. 1996). With the addition of the carbamazepine to the patient's regimen, there was an increased production of 3A4 over the ensuing weeks. Thus, more 3A4 was available to metabolize the methadone, resulting in a decrease (approximately 60%, according to Bell et al. [1988]) in the blood level of methadone over that time. This decrease in blood level led to an emerging opioid withdrawal syndrome and the need for an acute methadone dosage increase to address this. Discontinuation of the carbamazepine halted the increased production of 3A4. However, it takes weeks for the previously produced "extra" 3A4 to "die off" and for levels of 3A4 to return to baseline. This explains the need to gradually taper back to the original methadone dosage over the 2–3 weeks following the discontinuation of carbamazepine.

References

Bell J, Seres V, Bowron P, et al: The use of serum methadone levels in patients receiving methadone maintenance. Clin Pharmacol Ther 43(6):623–629, 1988

Iribarne C, Berthou F, Baird S, et al: Involvement of cytochrome P450 3A4 enzyme in the N-demethylation of methadone in human liver microsomes. Chem Res Toxicol 9(2):365–373, 1996

Spina E, Pisani F, Perucca E: Clinically significant pharmacokinetic drug interactions with carbamazepine: an update. Clin Pharmacokinet 31(3):198–214, 1996

Excessive Estrogen

A 33-year-old woman with recurrent anxious depression had done well on paroxetine (Paxil), 20 mg/day, for the past year. She also had been taking Loestrin, a specific oral contraceptive containing an especially small amount of ethinylestradiol (20 μg vs. the more standard 50 μg), because of an inability to tolerate more typical oral contraceptives. She had tolerated this regimen and had not become pregnant while regularly engaging in intercourse with her husband. However, she had been anorgasmic and moderately apathetic about sexual activity since beginning paroxetine therapy. After she consulted with her psychiatrist, her paroxetine was tapered and discontinued and nefazodone (Serzone) was immediately started. The nefazodone was titrated to a dosage of 400 mg/day. Although she encountered only mild sedation from the nefazodone, over the next 2 weeks she experienced increasing bloating, a 5-pound weight gain, and breast tenderness. These were the same symptoms she had experienced during her previous experiences with oral contraceptives containing higher amounts of ethinylestradiol. The nefazodone was discontinued and she was prescribed mirtazapine (Remeron), which she tolerated with only mild sedation. Her bloating, weight gain, and breast tenderness also abated with this change.

Discussion

This is an example of an inhibitor added to a substrate.

Ethinylestradiol is a 3A4 substrate (Guengerich 1990), and nefazodone is a strong 3A4 inhibitor (von Moltke et al. 1996). The addition of nefazodone impaired the efficiency of 3A4 in metabolizing the ethinylestradiol, resulting in an increase in the blood level of ethinylestradiol (Adson and Kotlyar 2001) to roughly the same level that the patient encountered during her previous experiences with oral contraceptives containing higher amounts of ethinylestradiol. This led to the rapid development of bloating, weight gain, and breast tenderness in this case. With the removal of the nefazodone, the blood level of ethinylestradiol returned to baseline and these symptoms remitted.

References

Adson DE, Kotlyar M: A probable interaction between a very low-dose oral contraceptive and the antidepressant nefazodone: a case report. J Clin Psychopharmacol 21:618–619, 2001

Guengerich FP: Metabolism of 17α-ethynylestradiol in humans. Life Sci 47(22): 1981–1988, 1990

von Moltke LL, Greenblatt DJ, Harmatz JS, et al: Triazolam biotransformation by human liver microsomes in vitro: effects of metabolic inhibitors and clinical confirmation of a predicted interaction with ketoconazole. J Pharmacol Exp Ther 276(2):370–379, 1996

Drowsy Dog Trainer

A 42-year-old woman who worked as a dog trainer was receiving alprazolam (Xanax), 1 mg tid, and citalopram (Celexa), 20 mg/day, from her psychiatrist for treatment of anxious depression. These medications had generally provided relief for her symptoms. During a life insurance physical examination with screening laboratory studies, she discovered that she was HIV-positive. Her internist prescribed, among other medications, the protease inhibitor ritonavir (Norvir). One week after starting the ritonavir, she felt so drowsy that she took a midafternoon nap, slept through her alarm, and therefore did not walk the dogs that she was scheduled to walk that day. Her disgruntled clients promptly fired her. Her despair over this event and her unremitting sedation led her to contact her psychiatrist. After reviewing recent developments, the psychiatrist changed her alprazolam to lorazepam (Ativan), 1 mg tid. The sedation remitted and she enjoyed a satisfactory clinical response.

Discussion

This is an example of an inhibitor added to a substrate.

Alprazolam is a 3A4 substrate (Dresser et al. 2000). Ritonavir is a potent 3A4 inhibitor (Iribarne et al. 1998), as well as a "pan-inhibitor" of all major P450 enzymes except for 1A2 and 2E1 (von Moltke et al. 1998). The addition of ritonavir to the patient's regimen impaired the ability of 3A4 to efficiently metabolize the alprazolam, resulting in an increase in the blood level of alprazolam despite no change in the dosing of the alprazolam (K.L. Cozza, personal communication, May 2002). This led to the patient's subsequent sedation. Her psychiatrist addressed this by changing from alprazolam to a benzodiazepine (lorazepam) that does not rely on P450 enzymes for its metabolism, but rather on phase II glucuronidation.

Since ritonavir becomes a specific *inducer* of 3A4 after a few weeks (Piscitelli et al. 2000), an alternative management strategy would have been to temporarily decrease the dosage of alprazolam to clinical effect when ritonavir's early 3A4 inhibitory effects predominated. A few weeks later, when ritonavir's specific 3A4 induction profile was able to offset its inhibitory profile, the alprazolam dosage could again be increased to clinical effect. The final alprazolam dosage could well be close to (or identical to) the original dosage before the addition of ritonavir, but there is no way to predict this. If anything, ritonavir's interaction with meperidine (Demerol) (see "Induction Toxicity," p. 63) suggests that ritonavir's 3A4 induction effect overtakes the inhibitory effect, which would suggest the need for a larger alprazolam final dosage. If one wished to adopt this strategy, decreasing and increasing the alprazolam dosage would need to be based on clinical response and side effects.

References

Dresser GK, Spence JD, Bailey DG: Pharmacokinetic-pharmacodynamic consequences and clinical relevance of cytochrome P450 3A4 inhibition. Clin Pharmacokinet 38(1):41–57, 2000

Iribarne C, Berthou F, Carlhant D, et al: Inhibition of methadone and buprenorphine N-dealkylations by three HIV-1 protease inhibitors. Drug Metab Dispos 26(3):257–260, 1998

Piscitelli SC, Kress DR, Bertz RJ, et al: The effect of ritonavir on the pharmacokinetics of meperidine and normeperidine. Pharmacotherapy 20(5):549–553, 2000

von Moltke LL, Greenblatt DJ, Grassi JM, et al: Protease inhibitors as inhibitors of human cytochromes P450: high risk associated with ritonavir. J Clin Pharmacol 38(2):106–111, 1998

Dyskinesias

A 14-year-old diagnosed with comorbid attention-deficit/hyperactivity disorder (ADHD) and Tourette's disorder was being successfully managed with methylphenidate (Ritalin SR), 20 mg bid, and risperidone (Risperdal), 2 mg bid. He then developed a seizure disorder, and a neurologist started him on carbamazepine, titrated to a dose of 300 mg bid (level = 8.2 μg/mL). Within 2 weeks of starting the carbamazepine, the patient developed new perioral dyskinesias (tongue thrusting, lip smacking, lip puckering) that were quite distinct from his Tourette's-related motor tics. The patient's carbamazepine was then discontinued, and he was started on divalproex sodium (Depakote). The dyskinesias persisted for roughly 10 days but then eventually remitted (J.R. Oesterheld, personal communication, May 2002).

Discussion

This is an example of an inducer added to a substrate.

Risperidone is a substrate of 2D6 and 3A4 (DeVane and Nemeroff 2001), and carbamazepine is a 3A4 inducer (Spina et al. 1996). The addition of carbamazepine led to a gradual increase in the production of 3A4 and therefore the amount of that enzyme available to metabolize the risperidone. Over the span of 2 weeks, this resulted in an acute two- to threefold decrease in the blood level of risperidone (deLeon and Bork 1997) and the subsequent emergence of withdrawal dyskinesias. Just as the presence of carbamazepine led to a gradual increase in the amount of 3A4, it took 1–2 weeks following the discontinuation of the carbamazepine before that "extra" 3A4 was no longer metabolically available and the amount of active 3A4 returned to baseline levels, resulting in a rise in the blood level of risperidone and a subsequent remission of the dyskinesias. It is also likely that the methylphenidate contributed to the development of the dyskinesias on a pharmacodynamic level.

References

deLeon J, Bork J: Risperidone and cytochrome P450 3A. J Clin Psychiatry 58(10): 450, 1997

DeVane CL, Nemeroff CB: An evaluation of risperidone drug interactions. J Clin Psychopharmacol 21(4):408–416, 2001

Spina E, Pisani F, Perucca E: Clinically significant pharmacokinetic drug interactions with carbamazepine: an update. Clin Pharmacokinet 31(3):198–214, 1996

Insomnia

A 28-year-old man with schizoaffective disorder had remained free of auditory hallucinations and paranoid delusions while taking a regimen of pimozide (Orap), 6 mg/day, and divalproex sodium (Depakote), 1,250 mg/

day. Over the past 2 months, he had developed progressive dysphoria, anergy, loss of appetite, and insomnia. He reported that he was sleeping only 4–5 hours each night and that this was making him "miserable." His psychiatrist decided to add nefazodone (Serzone), which was titrated to a dosage of 400 mg/day. Over the next week, the patient displayed severe restlessness and he reported the odd sensation that his heart was "fluttering." The psychiatrist discontinued the nefazodone, but he did check a stat ECG. The patient's QTc interval had elongated from 425 msec to 510 msec. Within 1 week of discontinuation of the nefazodone, the akathisia remitted and his ECG normalized.

Discussion

This is an example of an inhibitor added to a substrate.

Pimozide is a 3A4 substrate (Desta et al. 1999), and nefazodone is a strong 3A4 inhibitor (von Moltke et al. 1996). The addition of nefazodone impaired the ability of 3A4 to efficiently metabolize the pimozide, leading to an increase in the blood level of pimozide, even though the dosage had not been changed. This increased blood level of pimozide resulted in new-onset akathisia and QTc prolongation (K.L. Cozza, S.C. Armstrong, personal communication, May 2002) on his ECG. With the discontinuation of the nefazodone, 3A4 resumed its baseline (higher) level of activity, and this led to a return of the pimozide blood level back to its (lower) baseline, a remission of the akathisia, and a normalization of the ECG.

References

Desta Z, Kerbusch T, Flockhart DA: Effect of clarithromycin on the pharmacokinetics and pharmacodynamics of pimozide in healthy poor and extensive metabolizers of cytochrome P450 2D6 (CYP2D6). Clin Pharmacol Ther 65(1): 10–20, 1999

von Moltke LL, Greenblatt DJ, Harmatz JS, et al: Triazolam biotransformation by human liver microsomes in vitro: effects of metabolic inhibitors and clinical confirmation of a predicted interaction with ketoconazole. J Pharmacol Exp Ther 276(2):370–379, 1996

Legionnaire

A 47-year-old woman with generalized anxiety disorder (GAD) had responded well to buspirone (BuSpar), 30 mg bid, for several years. One winter, she developed a persistent dry cough, low-grade fever, persistent headache, and weakness. She visited her internist, who examined the patient and ran numerous tests before eventually diagnosing her with a Legionella pneumonia. He started the patient on erythromycin (E-mycin), 750 mg every 6 hours. Within 4 days, the patient's respiratory symptoms were significantly improved, but she then developed severe sedation, headache, nausea, and psychomotor slowing. She consulted her psychiatrist, who held her buspirone for 1 day and then instructed her to take buspirone, 5 mg bid, for the duration of her treatment with erythromycin. After her erythromycin course was completed, her buspirone dosage would be titrated back to 30 mg bid. She followed his instructions, and her treatment course proceeded uneventfully.

Discussion

This is an example of an inhibitor added to a substrate.

Buspirone is a 3A4 substrate (Kivisto et al. 1997), and erythromycin is a 3A4 inhibitor (Pai et al. 2000). The addition of the erythromycin impaired the efficient metabolism of buspirone by 3A4, leading to an increased blood level of buspirone despite an initially constant buspirone dosage. Some studies have indicated that the buspirone level may increase sixfold or more when buspirone is combined with erythromycin (Kivisto et al. 1997). The psychiatrist's suggested decrease in buspirone dosage compensated for this interaction. With the completion of the erythromycin course and the cessation of 3A4 inhibition, the buspirone dose could again return to 30 mg bid.

References

Kivisto KT, Lamberg TS, Kantola T, et al: Plasma buspirone concentrations are greatly increased by erythromycin and itraconazole. Clin Pharmacol Ther 62(3):348–354, 1997

Pai MP, Graci DM, Amsden GW: Macrolide drug interactions: an update. Ann Pharmacother 34(4):495–513, 2000

. . . Lub . . . Dub . . .

A 48-year-old lawyer diagnosed with anxious depression and hypertension had been maintained on citalopram (Celexa), 40 mg/day, and verapamil (Calan SR), 240 mg/day, for the past 2 years. However, he developed erectile dysfunction and chronic sleep difficulties on this regimen, and he met with his psychiatrist to change his antidepressant selection. After discussing the risks and benefits and the differing side-effect profiles, the patient and his psychiatrist decided to taper and discontinue the citalopram and begin a trial of nefazodone. The patient's nefazodone was titrated to a dosage of 500 mg/day over the course of 3 weeks. In the last week of this titration schedule, he experienced increasing dizziness when rising from a sitting or lying position. One week later, during an especially engaging episode of *Law & Order,* he abruptly passed out and his wife was unable to revive him. She called 911, and the EMS transported him to the nearest ER. During the ambulance ride, the heart monitor showed the presence of second-degree atrioventricular heart block with a ventricular rate of 41 beats per minute and a blood pressure of 70/30 mm Hg (Pies 2002).

Discussion

This is an example of an inhibitor added to a substrate.

Verapamil is a 3A4 substrate (Tracy et al. 1999), and nefazodone is a strong 3A4 inhibitor (von Moltke et al. 1996). The addition of the nefazodone impaired the ability of 3A4 to efficiently metabolize the verapamil (a calcium-channel blocker). This led to an increase in the blood level of verapamil, even though the dosage had not been changed. This state of verapamil toxicity created a dangerous condition in the form of second-degree atrioventricular heart block.

References

Pies R: Cytochromes and beyond: drug interactions in psychiatry. Psychiatric Times, May 2002, pp 48–51

Tracy TS, Korzekwa KR, Gonzalez FJ, et al: Cytochrome P450 isoforms involved in metabolism of the enantiomers of verapamil and norverapamil. Br J Clin Pharmacol 47(5):545–552, 1999

von Moltke LL, Greenblatt DJ, Harmatz JS, et al: Triazolam biotransformation by human liver microsomes in vitro: effects of metabolic inhibitors and clinical confirmation of a predicted interaction with ketoconazole. J Pharmacol Exp Ther 276(2):370–379, 1996

Renal Recklessness

A 37-year-old woman with no psychiatric history successfully received a renal transplant following a severe episode of rapidly progressive glomerulonephritis that was not recognized or treated in time. She was started on cyclosporine (Sandimmune), and she tolerated this medication without difficulty. She encountered no signs of organ rejection. However, her curtailment of her previously active lifestyle and body image issues led to increasing dysphoria, insomnia, poor concentration, and poor energy. Although a consulting psychiatrist recognized both psychological and medical contributions to her depressive symptoms, the patient strongly desired to start taking an antidepressant. The patient listed intact sexual functioning as an absolute priority and insomnia as her most troublesome symptom, which led the psychiatrist to select nefazodone (Serzone) as an appropriate antidepressant choice. Over the course of 4 weeks, he titrated the nefazodone to a dosage of 400 mg/day. Just as she reached this dosage, she began to grow increasingly lethargic and confused, becoming forgetful and losing track of the time, day, and date. She was normally vigilant about her health status, but in her impaired mental state she did not report her worsening oliguria to her nephrologist. She eventually had a tonic-clonic seizure, leading to immediate hospitalization, at which time she was found to be in acute renal failure (J. Sokal, personal communication, June 2002).

Discussion

This is an example of an inhibitor added to a substrate.

Cyclosporine is a 3A4 substrate (Kronbach et al. 1988), and nefazodone is a strong 3A4 inhibitor (von Moltke et al. 1996). When the nefazodone was added to the regimen, the ability of 3A4 to efficiently metabolize the cyclosporine was impaired. This led to an increase in the blood level of cyclosporine to the toxic range, even though the cyclosporine dosage had not been changed. As a result of her cyclosporine toxicity, the patient became acutely confused and lethargic, experienced a seizure, and developed acute renal failure. This interaction has also occurred when fluvoxamine (Luvox), a moderately strong 3A4 inhibitor, has been combined with cyclosporine (Vella and Sayegh 1998).

References

Kronbach T, Fischer V, Meyer UA: Cyclosporine metabolism in human liver: identification of a cytochrome P-450III gene family as the major cyclosporine-metabolizing enzyme explains interactions of cyclosporine with other drugs. Clin Pharmacol Ther 43(6):630–635, 1988

Vella JP, Sayegh MH: Interactions between cyclosporine and newer antidepressant medications. Am J Kidney Dis 31(2):320–323, 1998

von Moltke LL, Greenblatt DJ, Harmatz JS, et al: Triazolam biotransformation by human liver microsomes in vitro: effects of metabolic inhibitors and clinical confirmation of a predicted interaction with ketoconazole. J Pharmacol Exp Ther 276(2):370–379, 1996

Induction Toxicity

A 37-year-old, HIV-positive man, who was taking ritonavir (Norvir) as one of his chronic medications, was struck by a car while he was bicycling on the road. He was admitted to the nearest hospital and taken to the operating room, where he had an open reduction of his left fibula fracture. Postoperatively, he was transferred to the orthopedic floor and was given meperidine (Demerol), 25 mg im every 4 hours as needed, for postoperative pain. The meperidine was not providing effective analgesia, so in 2 days the dose was increased to 50 mg im every 4 hours as needed. Three days later, this dose provided only modest pain relief. The patient's visiting friends reported to the surgeon that he seemed more despondent and irritable than they would have expected, even under these circumstances. The patient was also noted to have developed a slight tremor. The next day, the patient had a grand mal seizure. After the patient recovered, the meperidine was immediately discontinued, and he was given morphine (MSIR) for his pain; the morphine was completely effective and produced none of the previous side effects.

Discussion

This is an example of a substrate added to an inducer.

Meperidine is at least a partial 3A4 substrate, and ritonavir is a 3A4 inducer when administered chronically (Piscitelli et al. 2000). Ritonavir is also a "pan-inhibitor" at all major P450 enzymes except for 1A2 and 2E1 (von Moltke et al. 1998). 3A4 catalyzes the specific transformation of meperidine into normeperidine (Piscitelli et al. 2000), which is a renally excreted toxic meperidine metabolite with a much longer half-life (15–20 hours) than that of the parent compound (3–4 hours) (Wong 2002). When the amount of metabolically available 3A4 is "normal," then the rate of renal elimination of normeperidine is generally sufficient to clear this compound and prevent toxic accumulations, so long as meperidine dosing remains moderate and the treatment course is relatively brief (several days). However, when ritonavir is chronically administered, the induction of 3A4 appears to overtake its inhibition. The resulting greater amount of available 3A4 leads to an increased rate of metabolic transformation of meperidine into the normeperidine metabolite, leading to decreased blood levels of meperidine and increased blood levels of normeperidine (normeperidine AUC is increased by 47%, according to Piscitelli et al. [2000]). In this case, the decreased meperidine led to less effective analgesia, while the increased normeperidine led to emerging dysphoria, irritability, tremor, and a seizure.

References

Piscitelli SC, Kress DR, Bertz RJ, et al: The effect of ritonavir on the pharmacokinetics of meperidine and normeperidine. Pharmacotherapy 20(5):549–553, 2000

von Moltke LL, Greenblatt DJ, Grassi JM, et al: Protease inhibitors as inhibitors of human cytochromes P450: high risk associated with ritonavir. J Clin Pharmacol 38(2):106–111, 1998

Wong: Wong on Web Paper: notes on meperidine. March 15, 2002. Available at: http://www.us.elsevierhealth.com/WOW/op024.html. Accessed July 29, 2002

Not Otherwise Specified

A 16-year-old carried the diagnoses of dissociative disorder not otherwise specified (NOS), impulse-control disorder NOS, and personality disorder NOS. She was a chronic self-mutilator; specifically, she regularly carved intricate designs into her left forearm with a razor. She was being treated with risperidone (Risperdal), 3 mg qhs, and topiramate (Topamax), 150 mg/day, and she appeared to tolerate this regimen without any notable side effects. The patient presented to a new clinic for an intake, and the new psychiatrist contacted the previous psychiatrist to inquire about the rationale for her medication regimen. She was told that the patient had a suboptimal but reasonably stable clinical course while taking the risperidone alone. The topiramate was added in pursuit of further improvements in impulse control, as well as to counteract a 10-pound weight gain that had occurred since starting the risperidone. The patient seemed to respond favorably for the first few weeks, but she then became more irritable and impulsive and began cutting herself at least twice each week. This decompensation was attributed to the stresses accompanying the start of the school year. Two months later, the patient's insurance coverage (via her mother) changed, requiring her to leave the previous psychiatrist and present for the clinic intake. This psychiatrist, who had extensive knowledge of P450 interactions, discontinued the topiramate, and the patient's irritability and self-cutting sharply decreased over the next 3 weeks (long before a therapeutic alliance had formed). One month later, the self-cutting had ceased altogether (J.R. Oesterheld, personal communication, August 2002).

Discussion

This is an example of both an inducer added to a substrate and reversal of induction.

Risperidone is a substrate of both 2D6 and 3A4 (DeVane and Nemeroff 2001), and topiramate is a 3A4 inducer (Benedetti 2000). With the initial addition of topiramate, more 3A4 was produced. This increased the efficiency with which 3A4 metabolized the risperidone, leading to a decreased risperidone blood level, even though the dosage of risperidone had not been changed. The decreased blood level of risperidone was at least partially responsible for the patient's subsequent clinical decompensation. With the eventual discontinuation of the topiramate, however, 3A4 returned to its lesser baseline amount and lower level of activity. This led to an increase in the risperidone blood level back to its pre-topiramate baseline, and a subsequent return of the clinical benefits that the risperidone had initially provided.

References

Benedetti MS: Enzyme induction and inhibition by new antiepileptic drugs: a review of human studies. Fundam Clin Pharmacol 14(4):301–319, 2000

DeVane CL, Nemeroff CB: An evaluation of risperidone drug interactions. J Clin Psychopharmacol 21(4):408–416, 2001

Stealing Sister's Sleepers

A 36-year-old, HIV-positive man with chronic depression had responded well to fluoxetine (Prozac), 20 mg/day, for the past 5 years. Several months ago, he began taking indinavir (Crixivan), 600 mg (vs. the typical 800 mg) every 8 hours, as one of his HIV medications. Over the past month, he was having more difficulties at his job, and his ruminations about these problems were leading to persistent onset insomnia. The patient lived with his parents and sister. One night when he was having trouble falling asleep, he sneaked into his sister's bathroom and took one of her 10-mg tablets of zolpidem (Ambien). He subsequently slept more than 15 hours and seemed "drugged" for the rest of the day, none of which was appreciated at his job (K.L. Cozza, personal communication, May 2002).

Discussion

This is an example of a substrate added to two inhibitors.

Zolpidem is principally a substrate of 3A4 (von Moltke et al. 1999). Indinavir is a strong 3A4 inhibitor (Iribarne et al. 1998), and fluoxetine's active metabolite, norfluoxetine, is a moderate 3A4 inhibitor (Greenblatt et al. 1999; von Moltke et al. 1995). The patient took a dose of zolpidem that would be expected to produce an average amount of sleep in a young man, with no residual daytime sedation. However, the presence of the dual 3A4 inhibitors very strongly interfered with 3A4's ability to efficiently metabolize the zolpidem. This led to a much higher blood level of zolpidem than would have been expected at this dose, with the result that the patient slept for an extended period and experienced substantial residual daytime sedation.

References

Greenblatt DJ, von Moltke LL, Harmatz JS, et al: Human cytochromes and some newer antidepressants: kinetics, metabolism, and drug interactions. J Clin Psychopharmacol 19 (5, suppl 1):23S–35S, 1999

Iribarne C, Berthou F, Carlhant D, et al: Inhibition of methadone and buprenorphine N-dealkylations by three HIV-1 protease inhibitors. Drug Metab Dispos 26(3):257–260, 1998

von Moltke LL, Greenblatt DJ, Court MH, et al: Inhibition of alprazolam and desipramine hydroxylation in vitro by paroxetine and fluvoxamine: comparison with other selective serotonin reuptake inhibitor antidepressants. J Clin Psychopharmacol 15(2):125–131, 1995

von Moltke LL, Greenblatt DJ, Granda BW, et al: Zolpidem metabolism in vitro: responsible cytochromes, chemical inhibitors, and in vivo correlations. Br J Clin Pharmacol 48(1):89–97, 1999

Unplanned Parenthood

A 35-year-old married woman with a seizure disorder had remained seizure-free for 5 years while taking carbamazepine (Tegretol). However, she was hoping to begin taking oral contraceptive medication, so she consulted with her neurologist. After discussing the risks and benefits of various agents, they decided on a trial of oxcarbazepine (Trileptal), partly because of the neurologist's belief that oxcarbazepine did not significantly induce the metabolism of other medications as carbamazepine did. She transitioned to oxcarbazepine without incident and then began an oral contraceptive that contained ethinylestradiol as the active ingredient. Much to her surprise, she was pregnant within the next year (J.R. Oesterheld, personal communication, May 2002).

Discussion

This is an example of a substrate added to an inducer.

Ethinylestradiol is a 3A4 substrate (Guengerich 1990), and oxcarbazepine is a 3A4 inducer (Wilbur and Ensom 2000), albeit not as powerful an inducer as carbamazepine. It is generally true that the increases in 3A4 induced by oxcarbazepine do not tend to exert clinically significant effects on most 3A4 substrates. However, it has been demonstrated that oxcarbazepine is capable of sufficient 3A4 induction to increase 3A4's efficiency in metabolizing ethinylestradiol to the point that ethinylestradiol-containing oral contraceptives significantly lose their efficacy (Elwes and Binnie 1996; Fattore et al. 1999). As with several other anticonvulsants, oxcarbazepine induces the glucuronidation of many other medications. Its ability to interact with other medications is not restricted to its induction of 3A4 (May et al. 1999).

References

Elwes RD, Binnie CD: Clinical pharmacokinetics of newer antiepileptic drugs: lamotrigine, vigabatrin, gabapentin and oxcarbazepine. Clin Pharmacokinet 30(6):403–415, 1996

Fattore C, Cipolla G, Gatti G, et al: Induction of ethinylestradiol and levonorgestrel metabolism by oxcarbazepine in healthy women. Epilepsia 40(6): 783–787, 1999

Guengerich FP: Metabolism of 17α-ethynylestradiol in humans. Life Sci 47(22): 1981–1988, 1990

May TW, Rambeck B, Jurgens U: Influence of oxcarbazepine and methsuximide on lamotrigine concentrations in epileptic patients with and without valproic acid comedication: results of a retrospective study. Ther Drug Monit 21(2): 175–181, 1999

Wilbur K, Ensom MH: Pharmacokinetic drug interactions between oral contraceptives and second-generation anticonvulsants. Clin Pharmacokinet 38(4): 355–365, 2000

Fungal Fatality

A 32-year-old, HIV-positive man with schizophrenia had been noncompliant with his pimozide (Orap), 4 mg/day, and his various HIV-related medications. He was eventually brought to a local ER by the police, after they found him standing in the middle of the street shouting and preaching at pedestrians and severely blocking traffic. He was admitted to the psychiatric inpatient unit, and his pimozide was restarted. He was noted to have a persistent cough and low-grade fevers. A workup eventually revealed that he had a pulmonary histoplasmosis infection. The consulting internist prescribed ketoconazole (Nizoral), 200 mg/day, to be given for at least 6 months. One week later, when the patient did not come to the unit kitchen for breakfast, staff went to the patient's room to wake him. He was not arousable. Although a code was called, it was clear that he had died several hours earlier (K.L. Cozza, personal communication, May 2002).

Discussion

This is an example of an inhibitor added to a substrate.

Pimozide is a 3A4 substrate (Desta et al. 1999), and ketoconazole is a very strong 3A4 inhibitor (Boxenbaum 1999). The addition of ketoconazole significantly impaired the ability of 3A4 to efficiently metabolize the pimozide. This led to a sharp increase in the blood level of pimozide, even though the dosage of pimozide had not been changed. Since pimozide prolongs the QTc interval on the ECG in a dose-dependent manner, the increase in the pimozide blood level caused a lengthening of the QTc interval, which likely led to the development of a torsades de pointes arrhythmia (Dresser et al. 2000), ventricular fibrillation, and then death.

References

Boxenbaum H: Cytochrome P450 3A4 in vivo ketoconazole competitive inhibition: determination of Ki and dangers associated with high clearance drugs in general. J Pharm Pharm Sci 2(2):47–52, 1999

Desta Z, Kerbusch T, Flockhart DA: Effect of clarithromycin on the pharmacokinetics and pharmacodynamics of pimozide in healthy poor and extensive metabolizers of cytochrome P450 2D6 (CYP2D6). Clin Pharmacol Ther 65(1): 10–20, 1999

Dresser GK, Spence JD, Bailey DG: Pharmacokinetic-pharmacodynamic consequences and clinical relevance of cytochrome P450 3A4 inhibition. Clin Pharmacokinet 38(1):41–57, 2000

Occupational Hazard

A 34-year-old sanitation worker with generalized anxiety disorder (GAD) was responding well to buspirone (BuSpar), 30 mg/day. One day, he picked up a plastic bag and two sharp, thick nails pierced through his gloves and into the base of the nail bed of his right index finger. Over the next 10 days, this area became extremely tender, red, and swollen, making it difficult for him to do his job. His primary care physician referred him to a dermatologist, who diagnosed a fungal infection (onychomycosis). The dermatologist prescribed itraconazole (Sporanox), 200 mg bid for 1 week, followed by a 3-week delay, then a repetition of the 200 mg bid for another week. By day 5 of the first week of itraconazole therapy, the patient had experienced marked sedation, nausea, and tremor. Even though he spent a lot of time sleeping, he was so highly motivated to treat this infection that he nonetheless complied with the itraconazole regimen. These symptoms persisted for almost a week after completion of the first week of itraconazole. When he reported these events to his psychiatrist, she instructed him to decrease his buspirone dosing by 5 mg/day until he reached a dosage of 5 mg/day. She instructed him to begin this buspirone taper on the first day of the second round of itraconazole. After the patient completed the itraconazole regimen, the psychiatrist then instructed the patient to increase his buspirone by 5 mg/day until he resumed his usual dosage of 30 mg/day. The patient followed these instructions faithfully, and during the second round of itraconazole he experienced only mild drowsiness, with no increase in his generalized anxiety.

Discussion

This is an example of an inhibitor added to a substrate.

Buspirone is a 3A4 substrate (Kivisto et al. 1997), and itraconazole is a strong 3A4 inhibitor (von Moltke et al. 1996). When the itraconazole was added, this significantly impaired the ability of 3A4 to efficiently metabolize the buspirone, resulting in a large increase in the blood level of buspirone, even though the buspirone dosage had remained constant during the first week of itraconazole therapy. This led to the symptoms of sedation, nausea, and tremor. When she was consulted, the psychiatrist anticipated this interaction and gave buspirone dosing instructions that successfully took into account itraconazole's half-life (64±32 hours) (Sporanox [package insert] 2002) and the magnitude of the interaction (itraconazole is reported to increase buspirone blood levels 5- to 13-fold) (Kivisto et al. 1997) for the second round of itraconazole therapy.

References

Sporanox (package insert). Montvale, NJ, Janssen Pharmaceutica, 2002

Kivisto KT, Lamberg TS, Kantola T, et al: Plasma buspirone concentrations are greatly increased by erythromycin and itraconazole. Clin Pharmacol Ther 62(3): 348–354, 1997 von Moltke LL, Greenblatt DJ, Schmider J, et al: Midazolam hydroxylation by human liver microsomes in vitro: inhibition by fluoxetine, norfluoxetine, and by azole antifungal agents. J Clin Pharmacol 36(9):783–791, 1996

A Weighty Matter

A 27-year-old woman with a history of bipolar I disorder and bulimia nervosa, after a recent psychiatric hospitalization, was taking olanzapine (Zyprexa), 10 mg qhs, and divalproex sodium (Depakote), 1,000 mg/day (blood level=95 μg/mL). She also was taking an oral contraceptive that contained ethinylestradiol. One month after discharge, she noticed that she had gained 10 pounds. She complained to her psychiatrist at length about this weight gain, ultimately threatening to stop all of her medications unless he addressed this issue to her satisfaction. The psychiatrist rather reluctantly added topiramate (Topamax), titrating up to a dosage of 200 mg/day. Although the patient did not lose weight, she did not gain any further weight. She did not experience any difficulties with cognition or paresthesias. Six weeks later, the patient was surprised to discover that she was having breakthrough bleeding at an irregular time in her cycle (J.R. Oesterheld, personal communication, May 2002).

Discussion

This is an example of an inducer added to a substrate.

Ethinylestradiol is a 3A4 substrate (Guengerich 1990), and topiramate is a 3A4 inducer (Benedetti 2000). The addition of topiramate led to an increase in the amount of 3A4 that was available to metabolize the ethinylestradiol, resulting in a decrease in the blood level of the ethinylestradiol, even though the dosage of ethinylestradiol had remained unchanged throughout. Topiramate has been found to decrease blood levels of ethinylestradiol by up to 30% (Rosenfeld et al. 1997). This decline in the ethinylestradiol component of the patient's oral contraceptive led to her episode of breakthrough bleeding.

References

Benedetti MS: Enzyme induction and inhibition by new antiepileptic drugs: a review of human studies. Fundam Clin Pharmacol 14(4):301–319, 2000

Guengerich FP: Metabolism of 17 α-ethynylestradiol in humans. Life Sci 47(22): 1981–1988, 1990

Rosenfeld WE, Doose DR, Walker SA, et al: Effect of topiramate on the pharmacokinetics of an oral contraceptive containing norethindrone and ethinyl estradiol in patients with epilepsy. Epilepsia 38(3):317–323, 1997

Patience

A 43-year-old, HIV-positive man, who was taking saquinavir (Fortovase), was having a bronchoscopy to evaluate some concerning masses noted on both a chest X ray and a computed tomography scan of the chest. He was given a standard dose (10 mg) of intravenous midazolam (Versed) throughout the course of the procedure. However, the patient did not awaken in the first hour after the procedure. The postoperative staff continued to wait for his sedation to abate. Finally, after having waited more than 5 hours postbronchoscopy, their patience was exhausted, at which time they administered flumazenil (Mazicon) to reverse his sedation (Merry et al. 1997). When he was discussing his extended period of sedation with one of the nurses, he remarked that he had always been "sensitive" to some medications. He recounted that roughly 10 years earlier (before he began taking saquinavir), he had tried one of his girlfriend's alprazolam (Xanax) pills one morning and had felt "really tired and spacey" for the rest of that day.

Discussion

This is an example of a substrate added to an inhibitor.

Midazolam is a 3A4 substrate (Dresser et al. 2000; von Moltke et al. 1996), and saquinavir is a mild- to moderate-strength 3A4 inhibitor (Iribarne et al. 1998). It also appears from his history that the patient may be an inefficient 3A4 metabolizer. This does not imply that he is a "poor metabolizer" at 3A4 (there are none such, as this would often be incompatible with life), but rather that he may be one of those individuals at the low end of the estimated 10- to 30-fold variability in the efficiency of 3A4 across the human species (Ketter et al. 1995). Therefore, since he was possibly a less active 3A4 metabolizer than most persons, he would likely have generated a higher midazolam blood level, and accordingly encountered more than typical sedation, following standard dosing of midazolam. However, with the addition of even a mild to moderate 3A4 inhibitor in this individual, who (possibly) had little 3A4 "reserve," the metabolism of midazolam was further slowed, and his midazolam blood level further increased, to the point that he eventually required the administration of a benzodiazepine antagonist to awaken.

References

Dresser GK, Spence JD, Bailey DG: Pharmacokinetic-pharmacodynamic consequences and clinical relevance of cytochrome P450 3A4 inhibition. Clin Pharmacokinet 38(1):41–57, 2000

Iribarne C, Berthou F, Carlhant D, et al: Inhibition of methadone and buprenorphine N-dealkylations by three HIV-1 protease inhibitors. Drug Metab Dispos 26(3):257–260, 1998

Ketter TA, Flockhart DA, Post RM, et al: The emerging role of cytochrome P450 3A in psychopharmacology. J Clin Psychopharmacol 15(6):387–398, 1995

Merry C, Mulcahy F, Barry M, et al: Saquinavir interaction with midazolam: pharmacokinetic considerations when prescribing protease inhibitors for patients with HIV disease. AIDS 11(2):268–269, 1997

von Moltke LL, Greenblatt DJ, Schmider J, et al: Midazolam hydroxylation by human liver microsomes in vitro: inhibition by fluoxetine, norfluoxetine, and by azole antifungal agents. J Clin Pharmacol 36(9):783–791, 1996

Stymied by Statins (II)

A 57-year-old man with hypercholesterolemia was taking simvastatin (Zocor) to address this. After his father passed away, he became increasingly anxious and depressed, reporting worsening terminal insomnia, anergy, poor appetite, and poor concentration to his internist. The internist had heard that nefazodone (Serzone) was helpful for anxious depression and especially helpful for insomnia. After some discussion with the patient, he prescribed nefazodone, titrating the drug to a dosage of 600 mg/day over the course of 3 weeks. In the week after reaching this dosage, the patient experienced worsening myalgias, muscle weakness, and fever, and his urine output declined sharply. He reported these symptoms to his internist, who advised the patient to go to an ER immediately. In the ER, he was found to have a CPK level over 2,000 IU/L and a serum creatinine level of 2.7 mg/dL. He was diagnosed with rhabdomyolysis-induced acute renal failure (S. Ruths, personal communication, June 2002).

Discussion

This is an example of an inhibitor added to a substrate.

Simvastatin is a 3A4 substrate (Gruer et al. 1999; Neuvonen et al. 1998), and nefazodone is a strong 3A4 inhibitor (von Moltke et al. 1996). The addition of nefazodone significantly impaired the ability of 3A4 to metabolize the simvastatin, which led to an increase in the blood level of simvastatin, even though the dosage had not changed. This resulted in a state of HMG-CoA reductase inhibitor toxicity, which in this case led to such extensive rhabdomyolysis that it overwhelmed the patient's renal function.

References

Gruer PJ, Vega JM, Mercuri MF, et al: Concomitant use of cytochrome P450 3A4 inhibitors and simvastatin. Am J Cardiol 84(7):811–815, 1999

Neuvonen PJ, Kantola T, Kivisto KT: Simvastatin but not pravastatin is very susceptible to interaction with the CYP3A4 inhibitor itraconazole. Clin Pharmacol Ther 63(3):332–341, 1998

von Moltke LL, Greenblatt DJ, Harmatz JS, et al: Triazolam biotransformation by human liver microsomes in vitro: effects of metabolic inhibitors and clinical confirmation of a predicted interaction with ketoconazole. J Pharmacol Exp Ther 276(2):370–379, 1996

Worry Wort

A 24-year-old woman with panic disorder and generalized anxiety disorder (GAD) had responded reasonably well to alprazolam (Xanax), 0.5 mg qid, prescribed by her primary care physician. After breaking up with her boyfriend of 3 years, she developed significant dysphoria and anhedonia. No other significant vegetative symptoms of depression were present. She was discussing her feelings of rejection and abandonment with her best friend, a devotee of homeopathy. The friend strongly suggested that she try St. John's wort *(Hypericum perforatum).* Without discussing it with her physician, she began to take the St. John's wort per the package instructions. In 2 weeks, she began to experience an increased frequency of panic attacks (one every other day), and within another week she was having daily attacks. Her generalized anxiety was similarly heightened. When she reported her difficulties to her physician, he referred the patient to a psychiatrist. After a thorough intake, the psychiatrist recommended discontinuing the St. John's wort, and he increased her standing alprazolam to 1 mg qid for 1 week, then 0.75 mg qid for 10 days, and then her baseline 0.5 mg qid thereafter. These instructions effectively prevented further panic attacks and treated her pervasively anxious state.

Discussion

This is an example of an inducer added to a substrate.

Alprazolam is a 3A4 substrate (Dresser et al. 2000), and St. John's wort is a 3A4 inducer (Moore et al. 2000; Roby et al. 2000). Addition of the St.

John's wort stimulated increased production of 3A4. The increase in active 3A4 led to increased metabolism of the alprazolam, which resulted in a decrease in the blood level of alprazolam over the course of 1–2 weeks (DeVane and Nemeroff 2002). This subtherapeutic blood level of alprazolam failed to prevent the recurrence of her panic attacks, which the psychiatrist compensated for by increasing the dosage of the substrate (alprazolam) and discontinuing the inducer (St. John's wort). Once the amount of 3A4 available to metabolize the alprazolam had returned to baseline levels, the patient was then able to resume her usual alprazolam dosing.

References

DeVane CL, Nemeroff CB: 2002 guide to psychotropic drug interactions. Primary Psychiatry 9(3):28–57, 2002

Dresser GK, Spence JD, Bailey DG: Pharmacokinetic-pharmacodynamic consequences and clinical relevance of cytochrome P450 3A4 inhibition. Clin Pharmacokinet 38(1):41–57, 2000

Moore LB, Goodwin B, Jones SA, et al: St. John's wort induces hepatic drug metabolism through activation of the pregnane X receptor. Proc Natl Acad Sci USA 97(13):7500–7502, 2000

Roby CA, Anderson GD, Kantor E, et al: St. John's wort: effect on CYP3A4 activity. Clin Pharmacol Ther 67(5):451–457, 2000

The Worst of Both Worlds

A 42-year-old woman who was taking carbamazepine (Tegretol), 1,000 mg/day (most recent level=11.1 µg/mL), for trigeminal neuralgia experienced a recurrence of her depression. She had responded well to nefazodone (Serzone), 500 mg/day, in the past, before she developed trigeminal neuralgia. Her psychiatrist again prescribed nefazodone, titrating to a dosage of 500 mg/day as before, but even after 6 weeks at that dosage there was no improvement in her depressive symptoms. A further increase to 600 mg/day yielded no further improvements. Additionally, the patient reported mildly increased fatigue and "forgetfulness" that she attributed to her depression, although the psychiatrist wondered if it was that simple. He ordered a carbamazepine blood level, which had risen to 14.1 µg/mL. He tapered and discontinued the nefazodone and started the patient on venlafaxine (Effexor). She eventually responded well to a dosage of 150 mg/day, and her fatigue and "forgetfulness" remitted even before the nefazodone taper was completed.

Discussion

This is a combined example of a substrate (nefazodone) added to an inducer (carbamazepine) and an inhibitor (nefazodone) added to a substrate (carbamazepine).

First, nefazodone, a 3A4 substrate (Serzone [package insert] 2001), was started at a dosage that previously had been effective. But the presence of carbamazepine, a 3A4 inducer (Spina et al. 1996), led to a greater amount of 3A4 than was present during the previous nefazodone trial and, hence, more extensive metabolism of the nefazodone during this trial. This resulted in a lower nefazodone blood level and an ineffective antidepressant trial. Studies have shown a greater than 10-fold decrease in nefazodone blood levels from baseline when nefazodone is coadministered with carbamazepine (Laroudie et al. 2000).

Second, when nefazodone, a 3A4 inhibitor (von Moltke et al. 1996), was added to carbamazepine, primarily a 3A4 substrate (although 1A2 and 2C9 make minor contributions to its metabolism) (Spina et al. 1996), the efficiency with which 3A4 was able to contribute to the metabolism of carbamazepine was impaired. Since the activity of 1A2 and 2C9 was not sufficient to compensate for nefazodone's inhibition of 3A4, there was a modest increase in the carbamazepine blood level (Laroudie et al. 2000) and the emergence of mild side effects (fatigue and mild confusion).

The decision to replace the nefazodone with venlafaxine, a 2D6 substrate (Effexor [package insert] 2002) that possesses only a mild 2D6 inhibitory profile (Ball et al. 1997), and whose metabolism is therefore not readily inducible by carbamazepine, led to antidepressant efficacy and a remission of the patient's mild carbamazepine toxicity.

References

Ball SE, Ahern D, Scatina J, et al: Venlafaxine: in vitro inhibition of CYP2D6 dependent imipramine and desipramine metabolism; comparative studies with selected SSRIs, and effects on human hepatic CYP3A4, CYP2C9 and CYP1A2. Br J Clin Pharmacol 43(6):619–626, 1997

Effexor (package insert). Philadelphia, PA, Wyeth Laboratories, April 2002

Laroudie C, Salazar DE, Cosson JP, et al: Carbamazepine-nefazodone interaction in healthy subjects. J Clin Psychopharmacol 20(1):46–53, 2000

Serzone (package insert). Princeton, NJ, Bristol-Myers Squibb, February 2001

Spina E, Pisani F, Perucca E: Clinically significant pharmacokinetic drug interactions with carbamazepine: an update. Clin Pharmacokinet 31(3):198–214, 1996

von Moltke LL, Greenblatt DJ, Harmatz JS, et al: Triazolam biotransformation by human liver microsomes in vitro: effects of metabolic inhibitors and clinical confirmation of a predicted interaction with ketoconazole. J Pharmacol Exp Ther 276(2):370–379, 1996

Clumsy

A 56-year-old woman with recurrent anxious depression had done well for several years on the regimen of mirtazapine (Remeron), 30 mg/day, and buspirone (BuSpar), 30 mg/day. During a routine checkup, her internist noted that her blood pressure had continued to steadily rise with age. After some discussion, he prescribed diltiazem (Cardizem) and scheduled a follow-up appointment in 1 month. However, within 3 days, the patient experienced growing lethargy and incoordination, which led to her frequently bumping into furniture and walls. After she reported these symptoms to the internist, he instructed her to discontinue the diltiazem and to come to his office that day. Her blood pressure was not unduly low, and she displayed no focal neurological signs. Although he was not aware of a specific interaction between diltiazem and her psychiatric medications, the chronology seemed to implicate this as the likely cause of her difficulties. He sent the patient home with instructions to check in with his office daily. Within 36 hours, she was feeling much better.

Discussion

This is an example of an inhibitor added to a substrate.

Buspirone is a 3A4 substrate (Kivisto et al. 1997), and diltiazem is a 3A4 inhibitor (Sutton et al. 1997). The addition of the diltiazem significantly impaired the ability of 3A4 to efficiently metabolize the buspirone. This led to an increase in the blood level of buspirone, even though the dosage of buspirone had not been changed, resulting in her lethargy and clumsiness. One study demonstrated a 5.5-fold increase in the buspirone AUC with the addition of diltiazem (Lamberg et al. 1998). Once the diltiazem was discontinued, 3A4 resumed its baseline (higher) level of metabolic activity, the buspirone blood level declined back to baseline, and the patient's side effects remitted.

References

Kivisto KT, Lamberg TS, Kantola T, et al: Plasma buspirone concentrations are greatly increased by erythromycin and itraconazole. Clin Pharmacol Ther 62(3): 348–354, 1997

Lamberg TS, Kivisto KT, Neuvonen PJ: Effects of verapamil and diltiazem on the pharmacokinetics and pharmacodynamics of buspirone. Clin Pharmacol Ther 63(6):640–645, 1998

Sutton D, Butler AM, Nadin L, et al: Role of CYP3A4 in human hepatic diltiazem N-demethylation: inhibition of CYP3A4 activity by oxidized diltiazem metabolites. J Pharmacol Exp Ther 282(1):294–300, 1997

Restless

A 32-year-old man who was receiving rifampin (in the form of Rifadin) as treatment for tuberculosis was having increasing difficulties with his relationships. He would often dwell on these as he tried to get to sleep, leading to persistent onset insomnia. He asked his internist for a sleep aid, and he was prescribed zolpidem (Ambien), 5–10 mg qhs prn for insomnia. The next night, he tried the 5-mg dose of zolpidem, but this did not help him get to sleep. The following night he tried 10 mg of zolpidem, and he still did not feel at all drowsy. Without consulting his internist, he continued to increase the dose. He found that only after he took a dose of at least 25 mg was he able to fall asleep promptly. After he reported these events to his internist, he was instead prescribed gabapentin (Neurontin), 300 mg po qhs as needed, for the insomnia, which did reliably help him to fall asleep.

Discussion

This is an example of a substrate added to an inducer.

Zolpidem is principally a substrate of 3A4 (von Moltke et al. 1999), and rifampin is a strong 3A4 inducer (Strayhorn et al. 1997). The zolpidem was initially prescribed at a dose that would generally help facilitate sleep. However, because rifampin was already present, there was much more 3A4 available to metabolize the zolpidem than was expected, with the result that the zolpidem blood level at a dose of 10 mg was far too low to promote sleep. Studies have shown that rifampin can reduce zolpidem's AUC to 27% of its baseline value and its maximum concentration to 42% of its baseline value (Villikka et al. 1997). Therefore, it was only after the patient increased the dose of the zolpidem, to compensate for the greater level of 3A4 activity due to induction by the rifampin, that he was able to fall asleep. Once the internist became aware of this situation, he switched to a sedating agent (the anticonvulsant gabapentin) that did not rely on P450-based metabolism, since rifampin is a "pan-inhibitor" across most major P450 enzymes.

References

Strayhorn VA, Baciewicz AM, Self TH: Update on rifampin drug interactions, III. Arch Intern Med 157(21):2453–2458, 1997

Villikka K, Kivisto KT, Luurila H, et al: Rifampin reduces plasma concentrations and effects of zolpidem. Clin Pharmacol Ther 62(6):629–634, 1997

von Moltke LL, Greenblatt DJ, Granda BW, et al: Zolpidem metabolism in vitro: responsible cytochromes, chemical inhibitors, and in vivo correlations. Br J Clin Pharmacol 48(1):89–97,1999

Oversuppression

A 62-year-old woman had undergone successful renal transplantation 6 months ago, and she was being stably maintained on the immunosuppressant regimen of prednisone and tacrolimus (Prograf), titrated to levels between 12 and 16 ng/mL. In the context of marital difficulties and numerous medical concerns, she was experiencing increasing anxiety, dysphoria, anhedonia, insomnia, poor appetite, and poor energy at that time. Her nephrologist decided to prescribe nefazodone (Serzone), 50 mg bid initially, with a plan to titrate to a dosage of 400 mg/day. However, after only 1 week of taking the initial 50 mg bid dosage of nefazodone, the patient experienced a coarse upper extremity tremor, confusion, and sedation. She was brought to the ER and admitted to the hospital. Her tacrolimus level had risen to 28.8 ng/mL, and her serum creatinine level had also risen, from 0.8 mg/dL to 1.9 mg/dL. Her nefazodone was then abruptly discontinued and the tacrolimus was held for 2 days. In less than 1 week, her tacrolimus level was again at baseline and her creatinine level was 1.0 mg/dL.

Discussion

This is an example of an inhibitor added to a substrate.

Tacrolimus is a 3A4 substrate (Olyaei et al. 1998), and nefazodone is a 3A4 inhibitor (von Moltke et al. 1996). When the nefazodone was added, the ability of 3A4 to efficiently metabolize the tacrolimus was significantly impaired, even at such a low dosage (50 mg bid) of nefazodone. This led to a higher blood level of tacrolimus, even though the tacrolimus dose had not been changed (Olyaei et al. 1998). This resulted in a state of tacrolimus toxicity, characterized by confusion (delirium), tremor, and sedation. After the nefazodone was removed, 3A4 then resumed its higher baseline level of activity, leading to a decrease of the tacrolimus blood level back to its baseline and a remission of her toxicity state.

There is also a likely contribution to this increase in the tacrolimus blood level from alterations in the functioning of the P-glycoprotein transporter. Tacrolimus is a P-glycoprotein substrate (Arima et al. 2001), and nefazodone is an acute P-glycoprotein inhibitor (when chronically administered, it also functions as a P-glycoprotein inducer) (Stormer et al. 2001). Thus, the nefazodone decreased the activity of P-glycoprotein, which led to less tacrolimus being extruded from enterocytes back into the gut lumen, where it would be excreted rather than absorbed. This resulting increase in the absorption of tacrolimus also contributed to the increase in the tacrolimus blood level following the addition of nefazodone.

References

Arima H, Yunomae K, Hirayama F, et al: Contribution of P-glycoprotein to the enhancing effects of dimethyl-beta-cyclodextrin on oral bioavailability of tacrolimus. J Pharmacol Exp Ther 297(2):547–555, 2001

Olyaei AJ, deMattos AM, Norman DJ, et al: Interaction between tacrolimus and nefazodone in a stable renal transplant recipient. Pharmacotherapy 18(6): 1356–1359, 1998

Stormer E, von Moltke LL, Perloff MD, et al: P-glycoprotein interactions of nefazodone and trazodone in cell culture. J Clin Pharmacol 41(7):708–714, 2001

von Moltke LL, Greenblatt DJ, Harmatz JS, et al: Triazolam biotransformation by human liver microsomes in vitro: effects of metabolic inhibitors and clinical confirmation of a predicted interaction with ketoconazole. J Pharmacol Exp Ther 276(2):370–379, 1996

Natural Disaster (II)

A 35-year-old, HIV-positive man was being maintained on nevirapine (Viramune), among other medications, with reasonably good control of his illness and good CD4 counts. Because of a recent breakup with his significant other, as well as the stresses of his HIV diagnosis, he began to experience increasing dysphoria, anhedonia, insomnia, and lack of energy. He independently decided to buy some St. John's wort *(Hypericum perforatum)* over the counter, and he began taking it as directed by the pharmacist. Over the next 3 months, he felt increasingly fatigued and developed a persistent cough. His cough grew painful, and he began to experience frequent fevers. He reported these symptoms to his internist, who instructed him to report to the nearest ER. Once in the ER, the patient had a chest X ray consistent with a diagnosis of *Pneumocystis carinii* pneumonia. Once the patient was admitted to the medical floor, it was found that his CD4 counts were significantly decreased and his viral load numbers were elevated compared with those obtained at his last office visit.

Discussion

This is an example of an inducer added to a substrate.

Nevirapine is a 3A4 substrate (Ioannides 2002), and St. John's wort is a 3A4 inducer (Moore et al. 2000; Roby et al. 2000). With the addition of the St. John's wort, the amount of 3A4 available to metabolize the nevirapine significantly increased, with the result that the blood level of nevirapine was significantly decreased (Ioannides 2002; "St. John's Wort Found to Lower Nevirapine Levels" 2001). In this case, the metabolism of the nevirapine was induced to the extent that the nevirapine blood level became subtherapeutic, and thus the nevirapine became ineffective at halting or slowing the progression of AIDS. Not only did the patient develop *P. carinii* pneumonia and worsening viral loads and CD4 counts, but exposing the patient in a sustained manner to a subtherapeutic nevirapine blood level could actually foster the development of resistant HIV strains, and thus decrease the future ability to treat the illness in this particular individual.

References

Ioannides C: Pharmacokinetic interactions between herbal remedies and medicinal drugs. Xenobiotica 32(6):451–478, 2002

Moore LB, Goodwin B, Jones SA, et al: St. John's wort induces hepatic drug metabolism through activation of the pregnane X receptor. Proc Natl Acad Sci USA 97(13):7500–7502, 2000

Roby CA, Anderson GD, Kantor E, et al: St. John's wort: effect on CYP3A4 activity. Clin Pharmacol Ther 67(5):451–457, 2000

St. John's wort found to lower nevirapine levels. Treatment Update 12(11):6, 2001

False Alarm

A 67-year-old man with a diagnosed seizure disorder had been successfully treated with carbamazepine in the past. However, there was some doubt as to whether he truly had an idiopathic seizure disorder or whether his seizures had resulted solely from prior episodes of alcohol withdrawal. As there had not been any seizures for more than 20 years, the patient and his neurologist decided that he could try to discontinue the carbamazepine, and indeed he was seizure-free for another 4 years while taking no antiepileptic medication. In the past 4 years, however, the patient did experience a transient ischemic attack (TIA) in which his initial symptom was slurred speech. His subsequent evaluation revealed significant hypertension (190/115 mm Hg), and he was placed on diltiazem (Cardizem SR), 180 mg bid, and aspirin, 325 mg/day. As a result of the diltiazem, his resulting blood pressure was 135/90 mm Hg. Unfortunately, he then had another seizure, and he and his neurologist decided that he should restart the carbamazepine. His prior dosage of carbamazepine had been 800 mg/day, which reliably produced blood levels between 7 and 10 μg/mL. Over the course of 2 weeks, his carbamazepine was titrated back to this dosage, which he initially tolerated without difficulty. However, by day 4 at this dosage of carbamazepine, he experienced more fatigue than he had expected, and when he woke up the next morning his speech was slurred. Fearing another TIA, he promptly called 911, and an ambulance transported him to the nearest ER. Once there, his neurologist evaluated him and ordered a carbamazepine blood level. His blood level was found to be 17.3 μg/mL, and a computed tomography scan of the brain was negative. His neurologist concluded that his slurred speech had been a result of mild carbamazepine toxicity and not another TIA.

Discussion

This is an example of a substrate added to an inhibitor.

Carbamazepine is primarily a 3A4 substrate, although 1A2 and 2C9 make minor contributions to the metabolism of carbamazepine (Spina et al. 1996). Diltiazem is a 3A4 inhibitor (Sutton et al. 1997). Carbamazepine was titrated to a dosage that had produced a particular blood level in the past. However, now that diltiazem was present, the ability of 3A4 to contribute to the efficient metabolism of carbamazepine was impaired. Since the activity of 1A2 and 2C9 was not sufficient to compensate for this effect, the presence of this "new" 3A4 inhibition resulted in a greater blood level of carbamazepine than when the patient had taken this dosage of carbamazepine in the past. Also, since full carbamazepine autoinduction was not likely to have occurred after 2 weeks of administration, the effect of diltiazem inhibition of 3A4 was maximal, but the counterbalancing effect of carbamazepine autoinduction had not yet reached its maximal ability to generate a lower baseline blood level. The addition of diltiazem to carbamazepine typically leads to carbamazepine blood level increases as high as 85% (Gadde and Calabrese 1990). The resulting state of mild carbamazepine toxicity accounted for the patient's fatigue and slurred speech.

References

Gadde K, Calabrese JR: Diltiazem effect on carbamazepine levels in manic depression. J Clin Psychopharmacol 10(5):378–379, 1990

Spina E, Pisani F, Perucca E: Clinically significant pharmacokinetic drug interactions with carbamazepine: an update. Clin Pharmacokinet 31(3):198–214, 1996

Sutton D, Butler AM, Nadin L, et al: Role of CYP3A4 in human hepatic diltiazem N-demethylation: inhibition of CYP3A4 activity by oxidized diltiazem metabolites. J Pharmacol Exp Ther 282(1):294–300, 1997

Cardiac Cross-Purposes

A 42-year-old woman with chronic paranoid schizophrenia had remained symptom-free for the past 3 years while taking quetiapine (Seroquel), 600 mg/day. During a routine examination with her primary care physician, her laboratory studies revealed that dietary interventions had not corrected her hypercholesterolemia. Her doctor decided to prescribe lovastatin (Mevacor), and he asked her to return in 3 months for a follow-up appointment. There were no clinical difficulties in the interim. When she came for the next appointment, an ECG and follow-up cholesterol level were obtained. Although her cholesterol level had declined, she also had a lengthening of the QTc interval on her ECG, from 443 msec to 497 msec. After consulting with the affiliated hospital's pharmacist, the doctor discontinued the lovastatin. A follow-up ECG 2 weeks later revealed that the QTc interval had returned to baseline.

Discussion

This is an example of an inhibitor added to a substrate.

Quetiapine is primarily a substrate of 3A4, with a minor contribution from 2D6 (DeVane and Nemeroff 2001). Lovastatin, while not typically considered a 3A4 inhibitor, is a sufficiently tightly bound substrate of 3A4 (Neuvonen and Jalava 1996) that it can potentially displace other 3A4 substrates from the active catalytic site of this enzyme, thus functioning as an inhibitor of 3A4's ability to metabolize other compounds. In other words, in this case, lovastatin is functioning as a clinically significant competitive inhibitor (as opposed to an allosteric inhibitor) with respect to quetiapine. Thus, the addition of the lovastatin likely impaired the ability of 3A4 to efficiently metabolize the quetiapine. This likely led to an increase in the blood level of quetiapine and the resulting increase in the QTc interval, which did reverse with the discontinuation of the lovastatin (Furst et al. 2002). Although quetiapine is not generally considered to present difficulties with cardiac function, the average QTc increase from baseline with quetiapine is 14.5 msec, with 11% of quetiapine-treated patients experiencing a QTc interval increase of at least 60 msec ("FDA's Safety Review of Ziprasidone [Zeldox]" 2000). These figures for QTc prolongation are lower than for thioridazine (Mellaril), mesoridazine (Serentil), or ziprasidone (Geodon), but they are higher than for most, if not all, of the other typical and atypical antipsychotic agents. This issue therefore bears some consideration when blood levels of quetiapine are elevated by coadministration with 3A4 inhibitors or when quetiapine is co-administered with other agents that prolong the QTc interval.

There is also an additional theoretical possibility that may explain this interaction. Lovastatin is a known inhibitor of P-glycoprotein (Bogman et al. 2001), and quetiapine is a known in vitro P-glycoprotein substrate (Boulton et al. 2002). This provides an additional mechanism by which lovastatin could elevate quetiapine blood levels, since lovastatin would impair the ability of P-glycoprotein to excrete quetiapine from enterocytes, leading to increased gastrointestinal absorption and thus increased quetiapine blood levels.

References

Bogman K, Peyer AK, Torok M, et al: HMG-CoA reductase inhibitors and P-glycoprotein modulation. Br J Pharmacol 132(6):1183–1192, 2001

Boulton DW, DeVane CL, Liston HL, et al: In vitro P-glycoprotein affinity for atypical and conventional antipsychotics. Life Sci 71(2):163–169, 2002

DeVane CL, Nemeroff CB: Clinical pharmacokinetics of quetiapine: an atypical antipsychotic. Clin Pharmacokinet 40(7):509–522, 2001

FDA's safety review of ziprasidone (Zeldox). Public Citizen's eLetter, pp 1–8, October 2000. Available at: http://www.citizen.org/ELETTER/ARTICLES/zeldox.htm. Accessed December 4, 2002

Furst BA, Champion KM, Pierre JM, et al: Possible association of QTc interval prolongation with co-administration of quetiapine and lovastatin. Biol Psychiatry 51(3):264–265, 2002

Neuvonen PJ, Jalava KM: Itraconazole drastically increases plasma concentrations of lovastatin and lovastatin acid. Clin Pharmacol Ther 60(1):54–61, 1996

Gradual Withdrawal (II)

A 19-year-old woman with panic disorder was responding well to alprazolam (Xanax), 1 mg tid, dispensed by her internist. She then experienced her first seizure, and she was referred to a neurologist. Her mother also had a chronic seizure disorder and had done quite well while taking carbamazepine (Tegretol). The neurologist suggested that the patient also start taking carbamazepine, and she agreed. The carbamazepine was titrated to a dosage of 800 mg/day (level=7.7 μg/mL) over a 2-week period, which the patient tolerated without difficulty. However, over the next week the patient experienced increasing anxiety and a return of her panic attacks. Additionally, she felt that her pulse was racing, and she experienced tremulousness and agitation. She reported her difficulties to her internist, who told her to report to the nearest ER. Her heart rate was found to be 116 beats per minute, and her blood pressure was 150/95 mm Hg. She was given lorazepam (Ativan), 2 mg po stat, and a prescription for 1 mg qid, with instructions to discontinue her alprazolam and to contact her internist immediately if any of these symptoms should recur.

Discussion

This is an example of an inducer added to a substrate.

Alprazolam is a 3A4 substrate (Dresser et al. 2000), and carbamazepine is a 3A4 inducer (Spina et al. 1996). The addition of the carbamazepine led to an increase in the amount of 3A4 that was available to metabolize the alprazolam. This resulted in a decrease in the blood level of alprazolam (Arana et al. 1988), to the extent that the patient actually entered into a state of mild benzodiazepine withdrawal. The astute ER physician addressed this difficulty by changing from alprazolam to lorazepam, a benzodiazepine that does not rely on P450 enzymes for its metabolism, and that also is not affected by carbamazepine's ability to induce specific glucuronidation enzymes.

References

Arana GW, Epstein S, Molloy M, et al: Carbamazepine-induced reduction of plasma alprazolam concentrations: a clinical case report. J Clin Psychiatry 49(11):448–449, 1988

Dresser GK, Spence JD, Bailey DG: Pharmacokinetic-pharmacodynamic consequences and clinical relevance of cytochrome P450 3A4 inhibition. Clin Pharmacokinet 38(1):41–57, 2000

Spina E, Pisani F, Perucca E: Clinically significant pharmacokinetic drug interactions with carbamazepine: an update. Clin Pharmacokinet 31(3):198–214, 1996

Inhibitor Induction

A 24-year-old, HIV-positive man was taking numerous medications for his illness, including indinavir (Crixivan), a protease inhibitor. An array of psychosocial stressors and ruminative somatic concerns led him to a state of increasing dysphoria, with accompanying insomnia, anxiety, poor concentration, and fatigue. He decided to buy some over-the-counter St. John's wort *(Hypericum perforatum)*, which he began to take per the package instructions. At his next appointment with his internist (3 months later), his CD4 counts had dropped precipitously. Further testing revealed both an increased viral load and a decreased indinavir blood level. When the internist asked about any recent life changes that might explain these findings, the patient then informed him about his decision to begin taking St. John's wort.

Discussion

This is an example of an inducer added to a substrate.

Indinavir is a 3A4 substrate (Barry et al. 1999), and St. John's wort is a 3A4 inducer (Moore et al. 2000; Roby et al. 2000). When the St. John's wort was added to the indinavir, this led to an increase in the amount of 3A4 that was available to metabolize the indinavir. As a result, there was a decrease in the indinavir blood level to the subtherapeutic range, with a subsequent decrease in the patient's CD4 count and an increase in his viral load. Studies have demonstrated that the addition of St. John's wort to indinavir can produce mean indinavir concentrations that are only 43% of baseline and trough indinavir blood levels that are only 19% of baseline (Piscitelli et al. 2000). Furthermore, protracted exposure to subtherapeutic concentrations of a protease inhibitor could predispose the patient to develop resistant strains of HIV.

There is also a likely P-glycoprotein contribution to this interaction. Indinavir is a P-glycoprotein substrate (van der Sandt et al. 2001), and St. John's wort is a P-glycoprotein inducer (Hennessy et al. 2002). Thus, the St. John's wort increased the activity of P-glycoprotein, which led to more indinavir being extruded from enterocytes back into the gut lumen, where it was excreted rather than absorbed. This increase in excretion and decrease in absorption of indinavir also likely contributed to the decrease in the blood level of indinavir following the addition of St. John's wort.

References

Barry M, Mulcahy F, Merry C, et al: Pharmacokinetics and potential interactions amongst antiretroviral agents used to treat patients with HIV infection. Clin Pharmacokinet 36(4):289–304, 1999

Hennessy M, Kelleher D, Spiers JP, et al: St Johns wort increases expression of P-glycoprotein: implications for drug interactions. Br J Clin Pharmacol 53(1): 75–82, 2002

Moore LB, Goodwin B, Jones SA, et al: St. John's wort induces hepatic drug metabolism through activation of the pregnane X receptor. Proc Natl Acad Sci USA 97(13):7500–7502, 2000

Piscitelli SC, Burstein AH, Chaitt D, et al: Indinavir concentrations and St. John's wort. Lancet 355(9203):547–548, 2000

Roby CA, Anderson GD, Kantor E, et al: St. John's wort: effect on CYP3A4 activity. Clin Pharmacol Ther 67(5):451–457, 2000

van der Sandt IC, Vos CM, Nabulsi L, et al: Assessment of active transport of HIV protease inhibitors in various cell lines and the in vitro blood-brain barrier. AIDS 15(4):483–491, 2001

Fungal Fatigue

A 54-year-old woman was receiving prednisone after successful liver transplantation. She was also taking zolpidem (Ambien), 10 mg qhs prn for insomnia, but she used this only about once per week. One month following the procedure, she began to develop a persistent cough and intermittent fevers. A readmission and workup revealed pulmonary histoplasmosis, for which ketoconazole was prescribed. One week later, she made use of her zolpidem prn, and she slept over 14 hours and felt extremely fatigued the next day. The pulmonologist discontinued the zolpidem and provided lorazepam (Ativan) prn for the insomnia.

Discussion

This is an example of a substrate added to an inhibitor.

Zolpidem is a 3A4 substrate (von Moltke et al. 1999), whereas ketoconazole is a strong 3A4 inhibitor (Boxenbaum 1999). Zolpidem was being dispensed at a standard hypnotic dose, given a typical level of 3A4 metabolic activity. However, when the ketoconazole was present, the ability of

3A4 to efficiently metabolize the zolpidem was markedly impaired. This led to a much greater blood level of zolpidem following the 10-mg dose than would have been expected had the ketoconazole not been present (DeVane and Nemeroff 2002), resulting in oversedation.

References

Boxenbaum H: Cytochrome P450 3A4 in vivo ketoconazole competitive inhibition: determination of Ki and dangers associated with high clearance drugs in general. J Pharm Pharm Sci 2(2):47–52, 1999

DeVane CL, Nemeroff CB: 2002 guide to psychotropic drug interactions. Primary Psychiatry 9(3):28–57, 2002

von Moltke LL, Greenblatt DJ, Granda BW, et al: Zolpidem metabolism in vitro: responsible cytochromes, chemical inhibitors, and in vivo correlations. Br J Clin Pharmacol 48(1):89–97, 1999

Compulsive Intoxication

A 27-year-old man with a history of opioid dependence was successfully participating in a methadone maintenance program, where he received methadone (Dolophine), 80 mg/day. However, he experienced a recurrence of his obsessive-compulsive disorder (OCD), characterized by repeated hand washing, checking, and hoarding behaviors, which derailed his usual ability to be socially interactive and hold down his job. His addiction specialist (an internist) decided to prescribe fluvoxamine (Luvox), having heard from various pharmaceutical representatives that this agent was specifically indicated for the treatment of OCD. Over the next 5 days, the patient progressively displayed the classic signs of opioid intoxication (miosis, sedation, slurred speech, and attentional deficits). The addiction specialist promptly discontinued the fluvoxamine, and these symptoms remitted over the succeeding 5 days. After consulting with a psychiatrist colleague, he then started the patient on citalopram (Celexa). This proved to be only partially effective in treating his OCD symptoms, even when titrated to a dosage of 60 mg/day, but there was no recurrence of opioid intoxication.

Discussion

This is an example of an inhibitor added to a substrate.

Fluvoxamine is a moderately strong 3A4 inhibitor and a strong 1A2, 2C9, and 2C19 inhibitor (Christensen et al. 2002; von Moltke et al. 1995). Methadone is a 3A4 substrate (Iribarne et al. 1996). The addition of the fluvoxamine significantly impaired the ability of 3A4 to efficiently metabolize the methadone. This led to an increase in the blood level of methadone, even though the dosage had not been changed. Studies have found that the addition of fluvoxamine to methadone can increase methadone blood levels by 20%–100% over baseline values (Bertschy et al. 1994). The increase in the blood level of methadone in this case was sufficient to generate a state of clinical opioid intoxication.

References

Bertschy G, Baumann P, Eap CB, et al: Probable metabolic interaction between methadone and fluvoxamine in addict patients. Ther Drug Monit 16(1):42–45, 1994

Christensen M, Tybring G, Mihara K, et al: Low daily 10-mg and 20-mg doses of fluvoxamine inhibit the metabolism of both caffeine (cytochrome P4501A2) and omeprazole (cytochrome P4502C19). Clin Pharmacol Ther 71(3):141–152, 2002

Iribarne C, Berthou F, Baird S, et al: Involvement of cytochrome P450 3A4 enzyme in the N-demethylation of methadone in human liver microsomes. Chem Res Toxicol 9(2):365–373, 1996

von Moltke LL, Greenblatt DJ, Court MH, et al: Inhibition of alprazolam and desipramine hydroxylation in vitro by paroxetine and fluvoxamine: comparison with other selective serotonin reuptake inhibitor antidepressants. J Clin Psychopharmacol 15(2):125–131, 1995

Cholesterol 451

A 58-year-old man with severe hypercholesterolemia had maintained a total cholesterol level of less than 250 mg/100 mL by taking atorvastatin (Lipitor). He then fell from a ladder and suffered a closed head injury. During the recovery period, he experienced a seizure and was placed on phenytoin (Dilantin) at a therapeutic dosage. Over the next month, he recovered fully from the sequelae of his fall, although he remained on the phenytoin for seizure prophylaxis. During his next routine visit with his internist, his cholesterol level had risen to 451 mg/100 mL.

Discussion

This is an example of an inducer added to a substrate.

Atorvastatin is a 3A4 substrate (Atorvastatin Monograf 2002), and phenytoin is a 3A4 inducer (phenytoin also induces 2C9 and 2C19) (Chetty et al. 1998; Spina and Perucca 2002). The addition of phenytoin led to increased production of 3A4 and, thus, more efficient metabolism of the atorvastatin. This resulted in a decrease in the blood level of atorvastatin, even though the dosage had not been changed (Atorvastatin Monograf 2002). This lower blood level of atorvastatin rendered it significantly less effective as a cholesterol-lowering agent via HMG-CoA reductase inhibition, which led to a significant increase in the serum cholesterol level.

References

Atorvastatin Monograf Ubat 180, pp 1–13, March 2002. Available at: http://www.bpfk.org/html/(180)%20Atorvastatin%20-%20March02.pdf. Accessed September 7, 2002

Chetty M, Miller R, Seymour MA: Phenytoin autoinduction. Ther Drug Monit 20(1):60–62, 1998

Spina E, Perucca E: Clinical significance of pharmacokinetic interactions between antiepileptic and psychotropic drugs. Epilepsia 43 (suppl 2):37–44, 2002

Sialorrhea

A 45-year-old woman with paranoid schizophrenia had experienced a much more stable clinical course since starting clozapine (Clozaril), 500 mg/day (blood level generally around 450 ng/mL). One winter, she developed a case of bronchitis and her internist prescribed erythromycin (E-mycin), 500 mg qid for 10 days. By day 5, she was experiencing some increased sedation, as well as constipation, and she began salivating profusely at night, requiring that she change pillows two to three times per night. She reported these symptoms to both her internist and her psychiatrist. Her psychiatrist ordered an immediate clozapine blood level, which had risen to 737 ng/mL. Her psychiatrist suggested that she take only 300 mg/day of clozapine until the day after her erythromycin course was completed and then resume her normal clozapine dosing. She followed his instructions, and the remainder of her course was uneventful, although no further blood levels were obtained at that time.

Discussion

This is an example of an inhibitor added to a substrate.

Clozapine's metabolism occurs primarily at 1A2. The enzymes 3A4 and 2C19 are also significant contributors, with 2D6 and 2C9 playing only minor metabolic roles (Eiermann et al. 1997; Olesen and Linnet 2001). Erythromycin, however, is a strong 3A4 inhibitor (Pai et al. 2000). The addition of erythromycin markedly impaired the ability of 3A4 to make a significant contribution to the overall metabolism of clozapine. Even though several P450 enzymes contribute to clozapine's overall metabolism, 3A4's role is sufficiently prominent that strong inhibition of this enzyme is able to reliably increase clozapine blood levels (Cohen et al. 1996). In this case, the increase in the clozapine blood level led to increased sedation and sialorrhea (hypersalivation). The effect of adding erythromycin to clozapine is variable in magnitude.

References

Cohen LG, Chesley S, Eugenio L, et al: Erythromycin-induced clozapine toxic reaction. Arch Intern Med 156(6):675–677, 1996

Eiermann B, Engel G, Johansson I, et al: The involvement of CYP1A2 and CYP3A4 in the metabolism of clozapine. Br J Clin Pharmacol 44(5):439–446, 1997

Olesen OV, Linnet K: Contributions of five human cytochrome P450 isoforms to the N-demethylation of clozapine in vitro at low and high concentrations. J Clin Pharmacol 41(8):823–832, 2001

Pai MP, Graci DM, Amsden GW: Macrolide drug interactions: an update. Ann Pharmacother 34(4):495–513, 2000

The Price of Vigilance

A 24-year-old medical student was recently diagnosed with narcolepsy. She happened to be on her third-year psychiatry rotation, and she asked her resident about the newest treatments for this condition. The resident mentioned modafinil (Provigil) and explained to her that this was felt to be "less addicting" than conventional psychostimulants. The student then approached her internist and requested that he prescribe modafinil for her. He agreed to try the patient on this medication, and he gradually titrated the drug to a dosage of 200 mg/day. She tolerated the modafinil well, and it did help with her daytime drowsiness and prevent "sleep attacks." However, 4 months later she discovered that she was pregnant, despite having faithfully taken oral contraceptive medication that contained ethinylestradiol.

Discussion

This is an example of an inducer added to a substrate.

Ethinylestradiol is a 3A4 substrate (Guengerich 1990), and modafinil is a 3A4 inducer (Provigil [package insert] 1999). With the addition of the modafinil, more 3A4 was produced and thus available to metabolize the ethinylestradiol. This led to a significant decrease in the blood level of ethinylestradiol, even though the dosage of ethinylestradiol had not been changed (Robertson and Hellriegel 2003). Without an effective blood level of the primary active ingredient (i.e., ethinylestradiol), her oral contraceptive medication became ineffective at preventing pregnancy.

References

Guengerich FP: Metabolism of 17α-ethynylestradiol in humans. Life Sci 47(22): 1981–1988, 1990

Provigil (package insert). West Chester, PA, Cephalon Inc, 1999

Robertson PJ, Hellriegel ET: Clinical pharmacokinetic profile of modafinil. Clin Pharmacokinet 42(2):123–137, 2003

Medicine Cabinet Misadventures

A 52-year-old woman with anxious depression had done quite well since being prescribed nefazodone (Serzone), 600 mg/day. In the winter of 1998, she developed a viral upper respiratory tract infection, characterized by a runny nose and sinus pressure leading to sinus headaches. It was a Sunday when she finally decided that she had to take something for relief, but the local pharmacy was closed. Before venturing into the cold to drive to a larger pharmacy, she decided to check the bathroom medicine cabinet for any medications that could be helpful. On her husband's side of the cabinet, she found an 18-month-old prescription for terfenadine (Seldane), of which 40 tablets still remained. She started taking one 60-mg tablet twice each day. She did not experience any significant relief with the terfenadine, but she decided to keep taking it and hope for the best. Six days after starting the terfenadine, she experienced a syncopal episode from which she did not spontaneously recover. Her husband immediately called 911, and the ambulance crew quickly arrived. When they attached a heart monitor, she was found to have a torsades de pointes arrhythmia. With quick stabilization and close monitoring on a telemetry unit, she eventually recovered completely. When the patient discussed these events with her internist, he explained to her that terfenadine had been taken off the market and why. He advised her to discard any remaining terfenadine and to call if she ever required an antihistamine, whereupon he would prescribe fexofenadine (Allegra).

Discussion

This is an example of a substrate added to an inhibitor.

This case has been included for historical interest to illustrate one of the more dramatic interactions that heightened awareness as to the importance of P450 issues in modern medicine. Terfenadine is a 3A4 substrate (Woosley 1996), and nefazodone is a strong 3A4 inhibitor (von Moltke et al. 1996). Actually, terfenadine is a pro-drug that relies on 3A4 to be transformed into fexofenadine, which is the active antihistaminic metabolite. Thus, in the presence of a strong 3A4 inhibitor like nefazodone, the ability of 3A4 to metabolize terfenadine into fexofenadine was significantly impaired, which led to both a lack of antihistaminic efficacy and an increase in the terfenadine blood level. Not only is the parent drug, terfenadine, ineffective as an antihistamine (until transformed to fexofenadine), it is also a cardiotoxic compound that prolongs the QTc interval with dangerous regularity (Renwick 1999). An increase in the levels of terfenadine by coadministration with a strong 3A4 inhibitor predisposed the patient to develop a torsades de pointes arrhythmia, which could have evolved into a frankly fatal arrhythmia (such as ventricular fibrillation). This propensity to develop torsades de pointes, and the accompanying patient morbidity and mortality, were the primary reasons that the antihistamines terfenadine and astemizole (Hismanal) and the promotility agent cisapride (Propulsid) were removed from the U.S. market within the past 6 years. Thioridazine (Mellaril), mesoridazine (Serentil), and pimozide (Orap) may well follow suit in the coming years.

References

Renwick AG: The metabolism of antihistamines and drug interactions: the role of cytochrome P450 enzymes. Clin Exp Allergy 29 (suppl 3):116–124, 1999

von Moltke LL, Greenblatt DJ, Harmatz JS, et al: Triazolam biotransformation by human liver microsomes in vitro: effects of metabolic inhibitors and clinical confirmation of a predicted interaction with ketoconazole. J Pharmacol Exp Ther 276(2):370–379, 1996

Woosley RL: Cardiac actions of antihistamines. Annu Rev Pharmacol Toxicol 36: 233–252, 1996

Avoidable Tragedy

A 47-year-old woman with recent sleep difficulties had been taking triazolam (Halcion), 0.25–0.5 mg qhs, prescribed by her family physician. She also developed a pedal onychomycosis, for which he prescribed itraconazole (Sporanox), 200 mg bid for the first week, to be followed by a 3-week hiatus, then another week of treatment. By day 5 of the first week of itraconazole use, the patient was experiencing significant daytime sedation, with slurring of her speech and ataxia. She contacted her neurologist, who advised her to hold her triazolam that night, and she was indeed more alert the next day. Despite these events, the patient nonetheless opted to take 0.375 mg of her triazolam that very night. She needed to make a 2-hour drive the following day. In the course of that drive, the patient fell deeply asleep at the wheel, crossing four lanes of traffic and the median barrier before crashing into a car heading the opposite direction, killing herself and both persons in the other vehicle (R. Love, personal communication, August 2002).

Discussion

This is an (especially tragic) example of an inhibitor added to a substrate.

Triazolam is a 3A4 substrate, as are all the other triazolobenzodiazepines, including midazolam (Versed) and alprazolam (Xanax) (Dresser et al. 2000). Itraconazole is a potent inhibitor of 3A4 (von Moltke et al. 1996). The addition of the itraconazole significantly impaired the ability of 3A4 to efficiently metabolize the triazolam, which led to a significant increase in the blood level and half-life of triazolam, even though the dose of triazolam had not been changed. It has been demonstrated that itraconazole can lengthen the half-life of triazolam from 1.5–5 hours (Hyman et al. 1995) to over 24 hours (Varhe et al. 1994). In this case, this effect accounted for the patient's severe daytime sedation and the subsequent tragedy.

References

Dresser GK, Spence JD, Bailey DG: Pharmacokinetic-pharmacodynamic consequences and clinical relevance of cytochrome P450 3A4 inhibition. Clin Pharmacokinet 38(1):41–57, 2000

Hyman SE, Arana GW, Rosenbaum JF: Handbook of Psychiatric Drug Therapy, 3rd Edition. Boston, MA, Little, Brown, 1995, p 151

Varhe A, Olkkola KT, Neuvonen PJ: Oral triazolam is potentially hazardous to patients receiving systemic antimycotics ketoconazole or itraconazole. Clin Pharmacol Ther 56(6, pt 1):601–607, 1994

von Moltke LL, Greenblatt DJ, Harmatz JS, et al: Triazolam biotransformation by human liver microsomes in vitro: effects of metabolic inhibitors and clinical confirmation of a predicted interaction with ketoconazole. J Pharmacol Exp Ther 276(2):370–379, 1996

The Spirit of Inquiry

A 28-year-old woman was receiving carbamazepine (Tegretol), 1,000 mg/day (blood level=9.3 µg/mL), for the treatment of bipolar I disorder. One winter, she contracted bronchitis, and her internist prescribed erythromycin (E-mycin), 500 mg qid for 10 days. The patient experienced increasing sedation and ataxia over the next several days, and on day 5 of this regimen she experienced a fall without loss of consciousness. She was taken to the nearest ER, where her carbamazepine blood level was found to be 20.9 µg/mL. She was admitted to the hospital, but no one contacted the prescribing psychiatrist, inquired as to her baseline carbamazepine blood levels at given doses, or even explored what led to the development of a toxic carbamazepine blood level. Rather, her carbamazepine was empirically decreased to 500 mg/day, and she was discharged from the hospital after 2 days of taking the carbamazepine at this dosage. She did not encounter any additional problems with the remainder of her erythromycin course. However, 5 days after the discontinuation of the erythromycin, the patient experienced the first and only seizure of her life. She was discovered by her boyfriend, who called an ambulance to again take her to the nearest ER, where her carbamazepine blood level was found to be 4.1 µg/mL (R. Love, personal communication, August 2002).

Discussion

This is an example of an inhibitor added to a substrate followed by reversal of inhibition.

Carbamazepine is primarily a 3A4 substrate, although 1A2 and 2C9 also make minor contributions to the metabolism of carbamazepine (Spina et al. 1996). Erythromycin is a 3A4 inhibitor (Pai et al. 2000). Initially, the addition of erythromycin impaired the ability of 3A4 to contribute to the efficient metabolism of carbamazepine. Since the activity of 1A2 and 2C9 was not sufficient to compensate for this effect, this inhibition of 3A4 led to an increase in the blood level of carbamazepine, even though the dosage of carbamazepine had not been increased. The resulting mild to moderate carbamazepine toxicity state manifested itself as sedation and ataxia, culminating in a fall. The hospitalist on the case compensated for the toxic carbamazepine blood level by decreasing the carbamazepine dosage, although he did not determine why and how this situation had developed. Subsequently, the hospitalist did not anticipate the reversal of inhibition that was to occur with the completion of the course of erythromycin. Once this occurred, 3A4 resumed its previous (higher) level of activity, which led to an increased rate of metabolism of carbamazepine and a subsequent decline in the blood level of carbamazepine. This rapid decline in the carbamazepine blood level led to a new-onset seizure. It was (and remains) unclear whether this seizure was entirely a product of her rapidly declining carbamazepine blood level, or whether these fluctuations in carbamazepine blood level served to unmask a seizure disorder that had been subclinical until that time. Whichever was the case, much morbidity could have been avoided if either the patient or the medical team had contacted the psychiatrist, who was well acquainted with the interaction of carbamazepine with erythromycin.

As with the case of "Ataxic" (see p. 47), a P-glycoprotein–mediated interaction may have further potentiated the P450 component of this interaction. Carbamazepine is a P-glycoprotein substrate (Potschka et al. 2001), and erythromycin is a P-glycoprotein inhibitor (Kiso et al. 2000). Thus, the addition of the erythromycin caused a decrease in the activity of P-glycoprotein in extruding carbamazepine from enterocytes back into the gut lumen, where it would be excreted rather than absorbed. This led to increased absorption of carbamazepine and thus an increase in the blood level of carbamazepine. The cessation of the erythromycin course led to a reversal of these P-glycoprotein–mediated effects, which then further contributed to the resulting decrease in the carbamazepine blood level.

References

Kiso S, Cai SH, Kitaichi K, et al: Inhibitory effect of erythromycin on P-glycoprotein–mediated biliary excretion of doxorubicin in rats. Anticancer Res 20(5A): 2827–2834, 2000

Pai MP, Graci DM, Amsden GW: Macrolide drug interactions: an update. Ann Pharmacother 34(4):495–513, 2000

Potschka H, Fedrowitz M, Loscher W: P-glycoprotein and multidrug resistance–associated protein are involved in the regulation of extracellular levels of the major antiepileptic drug carbamazepine in the brain. Neuroreport 12(16): 3557–3560, 2001

Spina E, Pisani F, Perucca E: Clinically significant pharmacokinetic drug interactions with carbamazepine: an update. Clin Pharmacokinet 31(3):198–214, 1996

Departed Decibels

A 65-year-old man with moderate gastroesophageal disease had been taking ranitidine (Zantac), 150 mg bid, for the past 2 years, with reasonably good results. One winter, he developed persistent cough, fevers, occasional shortness of breath, and pleuritic chest pain. He presented to a local ER, where a chest X ray revealed an atypical pneumonia. He was started on oral amoxicillin (Trimox), 1,000 mg bid, and erythromycin (E-mycin), 500 mg qid. Four days later, the patient requested and received a prescription for cimetidine (Tagamet), 300 mg qid, in the place of his previous ranitidine for cost reasons. Over the course of the next 5 days, the patient's pulmonary symptoms improved, but he complained about progressive "fuzzy hearing." He was eventually referred to an ENT specialist, who determined that the patient had likely suffered a bilateral 40- to 60-decibel hearing loss (J.R. Oesterheld, personal communication, July 2002).

Discussion

This is an example of an inhibitor added to a substrate.

Erythromycin is a 3A4 substrate (Erythromycin [package insert] 2000), and cimetidine is a "pan-inhibitor" that strongly inhibits 3A4, 2D6, and 1A2 (Martinez et al. 1999). Thus, the addition of the cimetidine significantly impaired the ability of 3A4 to metabolize erythromycin. This led to an increase in the blood level of erythromycin, which led to erythromycin-induced ototoxicity.

References

Erythromycin (package insert). North Chicago, IL, Abbott Laboratories, 2000

Martinez C, Albet C, Agundez JA, et al: Comparative in vitro and in vivo inhibition of cytochrome P450 CYP1A2, CYP2D6, and CYP3A by H2-receptor antagonists. Clin Pharmacol Ther 65(4):369–376, 1999

Insufficiency

A 42-year-old man had successfully undergone liver transplantation, following which he was placed on standing prednisone (Deltasone) as an immunosuppressant. He also was prescribed standing fluconazole (Diflucan) as a prophylactic antifungal agent. His fluconazole was eventually discontinued, but he remained on prednisone. Within the next 5 days, he experienced acute onset of fever, vomiting, abdominal pain, delirium, and hypotension. He was quickly given supplementary intravenous cortisol, and an endocrinologist was consulted to address his sudden-onset adrenal insufficiency (Tiao et al. 1999).

Discussion

This is an example of reversal of inhibition.

Prednisone is a 3A4 substrate (Schwab and Klotz 2001), and fluconazole is a strong 2C9 inhibitor and a moderate 3A4 inhibitor (Michalets and Williams 2000; Niemi et al. 2001). The presence of the fluconazole significantly impaired the ability of 3A4 to metabolize the prednisone. This led to a higher blood level of prednisone than would have existed had the fluconazole not been a part of the patient's regimen. When the fluconazole was discontinued, 3A4 returned to its baseline (higher) level of functioning. This led to more efficient metabolism of prednisone by 3A4, with the result that the blood level of prednisone then precipitously declined. This caused the patient to experience an acute addisonian crisis, requiring emergent administration of additional intravenous cortisol to avert a potential disaster.

References

Michalets EL, Williams CR: Drug interactions with cisapride: clinical implications. Clin Pharmacokinet 39(1):49–75, 2000

Niemi M, Backman JT, Neuvonen M, et al: Effects of fluconazole and fluvoxamine on the pharmacokinetics and pharmacodynamics of glimepiride. Clin Pharmacol Ther 69(4):194–200, 2001

Schwab M, Klotz U: Pharmacokinetic considerations in the treatment of inflammatory bowel disease. Clin Pharmacokinet 40(10):723–751, 2001

Tiao GM, Martin J, Weber FL, et al: Addisonian crisis in a liver transplant patient due to fluconazole withdrawal. Clin Transplant 13 (1, pt 1):62–64, 1999

Anticonvulsant Withdrawal Intoxication

A 28-year-old man with a history of opioid dependence and a seizure disorder was receiving phenytoin (Dilantin), 400 mg/day (blood level = 16.3 μg/mL), from his neurologist and methadone (Dolophine), 80 mg/day, from the methadone clinic at which he was enrolled. He had been doing well, but he began to develop gum hyperplasia and enlargement of his lips. After consulting with his neurologist, he was crossed over from the phenytoin to divalproex sodium (Depakote). In the weeks during which the phenytoin was tapered, the patient experienced increasing sedation and incoordination. Within 3 days of the discontinuation of the phenytoin, he felt frankly "high" and reported this to his neurologist and his counselor at the methadone clinic. After some discussion, the psychiatrist at the methadone clinic decreased his methadone dosage to 40 mg/day, which provided relief from his cravings and a cessation of his intoxicated state.

Discussion

This is an example of reversal of induction.

Methadone is substrate of 3A4 (Iribarne et al. 1996), and phenytoin is an inducer of 3A4, 2C9, and 2C19 (Chetty et al. 1998; Spina and Perucca 2002). The original appropriate dosage of methadone was influenced by the presence of the phenytoin, which increased the amount of 3A4 that was available to metabolize the methadone. This necessitated a higher dosage of methadone than would have been needed if the phenytoin had not been present (Tong et al. 1981). Thus, with the taper and discontinuation of the phenytoin, the amount of 3A4 decreased over the succeeding 2–3 weeks, which returned 3A4 to its lower baseline level of activity. This led to an increase in the blood level of methadone, even though the methadone dosage had not been increased. The clinic psychiatrist compensated for this reversal of induction by halving the methadone dosage, which provided therapeutic efficacy and a remission of intoxication.

References

Chetty M, Miller R, Seymour MA: Phenytoin autoinduction. Ther Drug Monit 20(1):60–62, 1998

Iribarne C, Berthou F, Baird S, et al: Involvement of cytochrome P450 3A4 enzyme in the N-demethylation of methadone in human liver microsomes. Chem Res Toxicol 9(2):365–373, 1996

Spina E, Perucca E: Clinical significance of pharmacokinetic interactions between antiepileptic and psychotropic drugs. Epilepsia 43 (suppl 2):37–44, 2002

Tong TG, Pond SM, Kreek MJ, et al: Phenytoin-induced methadone withdrawal. Ann Intern Med 94(3):349–351, 1981

Natural Disaster (III)

A 58-year-old man had just undergone a successful renal transplantation procedure. He was being maintained on tacrolimus (Prograf), titrated to a level of roughly 14 ng/mL, and prednisone. He began to experience some depressive symptoms in the weeks after the operation. When he had felt "blue" in the past, he took St. John's wort *(Hypericum perforatum)*, and he experienced it as helpful. He asked his wife to pick up some St. John's wort at the supermarket pharmacy, and he resumed taking this agent. Roughly 1 month later, the patient developed oliguria, tenderness around the graft site, and intermittent fevers. He contacted his nephrologist, who advised him to report to the nearest ER, where he was admitted and a workup revealed that he was in the acute stage of organ rejection (S.C. Armstrong, personal communication, May 2002).

Discussion

This is an example of an inducer added to a substrate.

Tacrolimus is a 3A4 substrate (Olyaei et al. 1998), and St. John's wort is a 3A4 inducer (Moore et al. 2000; Roby et al. 2000). With the addition of the St. John's wort, more 3A4 was produced and thus available to metabolize the tacrolimus. This led to a decrease in the blood level of tacrolimus, even though there had been no decrease in the tacrolimus dosage (Bolley et al. 2002). This decreased tacrolimus blood level resulted in less effective immunosuppression, which allowed the body's immune system to reject the new kidney.

There was also a likely contribution to this decrease in the tacrolimus blood level from alterations in the functioning of the P-glycoprotein transporter. Tacrolimus is a P-glycoprotein substrate (Arima et al. 2001), and St. John's wort is a P-glycoprotein inducer (Hennessy et al. 2002). Thus, the St. John's wort increased the activity of P-glycoprotein, which led to more tacrolimus being extruded from enterocytes back into the gut lumen, where it was excreted rather than absorbed. This increase in excretion and decrease in absorption of tacrolimus also contributed to the decrease in the tacrolimus blood level following the addition of St. John's wort.

References

Arima H, Yunomae K, Hirayama F, et al: Contribution of P-glycoprotein to the enhancing effects of dimethyl-beta-cyclodextrin on oral bioavailability of tacrolimus. J Pharmacol Exp Ther 297(2):547–555, 2001

Bolley R, Zulke C, Kammerl M, et al: Tacrolimus-induced nephrotoxicity unmasked by induction of the CYP3A4 system with St. John's wort (letter). Transplantation 73(6):1009, 2002

Hennessy M, Kelleher D, Spiers JP, et al: St Johns wort increases expression of P-glycoprotein: implications for drug interactions. Br J Clin Pharmacol 53(1): 75–82, 2002

Moore LB, Goodwin B, Jones SA, et al: St. John's wort induces hepatic drug metabolism through activation of the pregnane X receptor. Proc Natl Acad Sci USA 97(13):7500–7502, 2000

Olyaei AJ, deMattos AM, Norman DJ, et al: Interaction between tacrolimus and nefazodone in a stable renal transplant recipient. Pharmacotherapy 18(6): 1356–1359, 1998

Roby CA, Anderson GD, Kantor E, et al: St. John's wort: effect on CYP3A4 activity. Clin Pharmacol Ther 67(5):451–457, 2000

Chapter 4

1A2 and Complex Clozapine/ Olanzapine Case Vignettes

The 1A2 enzyme constitutes 13% of the liver's P450 complement, and it is found only in the liver (DeVane and Nemeroff 2002). Common 1A2 substrates include tricyclic antidepressants (TCAs) (Sawada and Ohtani 2001; Zhang and Kaminsky 1995), clozapine (Eiermann et al. 1997), olanzapine (Callaghan et al. 1999), and xanthines (Miners and Birkett 1996). Common 1A2 inhibitors include quinolones (Batty et al. 1995), fluvoxamine (Brosen 1995), some antiarrhythmic agents (Kobayashi et al. 1998; Nakajima et al. 1998), and grapefruit juice (Fuhr 1998). Common 1A2 inducers include tobacco smoking (Schrenk et al. 1998; Zevin and Benowitz 1999), charred meats and several cruciferous vegetables (Jefferson 1998; Jefferson and Greist 1996; Kall et al. 1996), carbamazepine (Parker et al. 1998), and rifampin (Wietholtz et al. 1995).

Polymorphisms for 1A2 do exist. Alam and Sharma (2001) identified three distinct 1A2 alleles besides the wild type.

References

Alam DA, Sharma RP: Cytochrome enzyme genotype and the prediction of therapeutic response to psychotropics. Psychiatric Annals 31(12):715–722, 2001

Batty KT, Davis TM, Ilett KF, et al: The effect of ciprofloxacin on theophylline pharmacokinetics in healthy subjects. Br J Clin Pharmacol 39(3):305–311, 1995

Brosen K: Drug interactions and the cytochrome P450 system: the role of cytochrome P450 1A2. Clin Pharmacokinet 29 (suppl 1):20–25, 1995

Callaghan JT, Bergstrom RF, Ptak LR, et al: Olanzapine: pharmacokinetic and pharmacodynamic profile. Clin Pharmacokinet 37(3):177–193, 1999

DeVane CL, Nemeroff CB: 2002 guide to psychotropic drug interactions. Primary Psychiatry 9(3):28–57, 2002

Eiermann B, Engel G, Johansson I, et al: The involvement of CYP1A2 and CYP3A4 in the metabolism of clozapine. Br J Clin Pharmacol 44(5):439–446, 1997

Fuhr U: Drug interactions with grapefruit juice: extent, probable mechanism and clinical relevance. Drug Saf 18(4):251–272, 1998

Jefferson JW: Drug and diet interactions: avoiding therapeutic paralysis. J Clin Psychiatry 59 (suppl 16):31–39, 40–42, 1998

Jefferson JW, Greist JH: Brussels sprouts and psychopharmacology: understanding the cytochrome P450 enzyme system. The Psychiatric Clinics of North America Annual of Drug Therapy 3:205–222, 1996

Kall MA, Vang O, Clausen J: Effects of dietary broccoli on human in vivo drug metabolizing enzymes: evaluation of caffeine, oestrone and chlorzoxazone metabolism. Carcinogenesis 17(4):793–799, 1996

Kobayashi K, Nakajima M, Chiba K, et al: Inhibitory effects of antiarrhythmic drugs on phenacetin O-deethylation catalysed by human CYP1A2. Br J Clin Pharmacol 45(4):361–368, 1998

Miners JO, Birkett DJ: The use of caffeine as a metabolic probe for human drug metabolizing enzymes. Gen Pharmacol 27(2):245–249, 1996

Nakajima M, Kobayashi K, Shimada N, et al: Involvement of CYP1A2 in mexiletine metabolism. Br J Clin Pharmacol 46(1):55–62, 1998

Parker AC, Pritchard P, Preston T, et al: Induction of CYP1A2 activity by carbamazepine in children using the caffeine breath test. Br J Clin Pharmacol 45(2):176–178, 1998

Sawada Y, Ohtani H: Pharmacokinetics and drug interactions of antidepressive agents (in Japanese). Nippon Rinsho 59(8):1539–1545, 2001

Schrenk D, Brockmeier D, Morike K, et al: A distribution study of CYP1A2 phenotypes among smokers and non-smokers in a cohort of healthy Caucasian volunteers. Eur J Clin Pharmacol 53(5):361–367, 1998

Wietholtz H, Zysset T, Marschall HU, et al: The influence of rifampin treatment on caffeine clearance in healthy man. J Hepatol 22(1):78–81, 1995

Zevin S, Benowitz NL: Drug interactions with tobacco smoking: an update. Clin Pharmacokinet 36(6):425–438, 1999

Zhang ZY, Kaminsky LS: Characterization of human cytochromes P450 involved in theophylline 8-hydroxylation. Biochem Pharmacol 50(2):205–211, 1995

Short-Term Gains

A 42-year-old man with schizoaffective disorder, with a previously difficult clinical course, had not required hospitalization since he was started on clozapine (Clozaril), 300 mg qhs (blood level=691 ng/mL), roughly 3 years ago. He was taking pantoprazole (Protonix), 20 mg/day, for his reflux esophagitis, but his insurance plan co-pay for this medication climbed dramatically. He consulted his internist, who prescribed cimetidine (Tagamet), 300 mg qid, in an effort to save the patient some money. However, within 5 days, the patient experienced increasing sedation, ataxia, and dizziness, culminating in a syncopal episode that required a medical admission to rule out cardiac causes for his syncope. However, in the hospital, his clozapine blood level was found to be 1,552 ng/mL.

Discussion

This is an example of an inhibitor added to a substrate.

Clozapine's metabolism occurs primarily at 1A2. The enzymes 3A4 and 2C19 are also significant contributors, with 2D6 and 2C9 playing only minor metabolic roles (Eiermann et al. 1997; Olesen and Linnet 2001). Cimetidine is a "pan-inhibitor" of 1A2, 3A4, 2D6, 2C9, and 2C19 (Martinez et al. 1999; Nation et al. 1990). Thus, the addition of cimetidine impaired the efficiency of all the P450 enzymes that play any significant role in the metabolism of clozapine. This decrease in the metabolism of clozapine led to a significant rise in the clozapine blood level (Szymanski et al. 1991) (more than double in this case), even though the clozapine dosage had not been changed. This resulted in the patient's sedation, orthostasis, and syncope.

References

Eiermann B, Engel G, Johansson I, et al: The involvement of CYP1A2 and CYP3A4 in the metabolism of clozapine. Br J Clin Pharmacol 44(5):439–446, 1997

Martinez C, Albet C, Agundez JA, et al: Comparative in vitro and in vivo inhibition of cytochrome P450 CYP1A2, CYP2D6, and CYP3A by H2-receptor antagonists. Clin Pharmacol Ther 65(4):369–376, 1999

Nation RL, Evans AM, Milne RW: Pharmacokinetic drug interactions with phenytoin, Part I. Clin Pharmacokinet 18(1):37–60, 1990

Olesen OV, Linnet K: Contributions of five human cytochrome P450 isoforms to the N-demethylation of clozapine in vitro at low and high concentrations. J Clin Pharmacol 41(8):823–832, 2001

Szymanski S, Lieberman JA, Picou D, et al: A case-report of cimetidine-induced clozapine toxicity. J Clin Psychiatry 52(1):21–22, 1991

VIP Psychosis (I)

A 32-year-old woman with chronic paranoid schizophrenia was being stably maintained on clozapine (Clozaril), 500 mg/day (blood level=457 ng/mL). Her affluent family insisted on taking her with them on an

African safari, despite the misgivings of her psychiatrist. On several occasions she wandered off from their hotel, and her family eventually discovered that she was engaged in a physically intimate relationship with one of the bellhops, who resided in a nearby village. Her parents whisked her away, and they all returned to the United States soon thereafter. Over the next several weeks, the patient developed progressive cough, chills, fatigue, and intermittent fevers. After an extensive workup, she was diagnosed with tuberculosis. The parents insisted on taking the patient to the most prominent infectious disease specialist in the area, although he was semiretired at the time. He placed the patient on rifampin (in the form of Rifadin), among other medications. The parents did not spontaneously disclose that the patient was taking clozapine, as they were accustomed to downplaying their daughter's mental illness, and the infectious disease specialist did not obtain as rigorous a history as was customary, in an effort to accommodate his important client(s). Thus, he remained unaware of this fact. Within 1 month of starting rifampin, the patient was floridly psychotic and hypersexual. She was initially admitted to a local day hospital, but she eventually required an admission to the inpatient unit after attempting to engage in inappropriate sexual activity with a male fellow patient on the front lawn of the day hospital. Her clozapine blood level was then discovered to be only 92 ng/mL.

Discussion

This is an example of an inducer added to a substrate.

Clozapine's metabolism occurs primarily at 1A2. The enzymes 3A4 and 2C19 are also significant contributors, with 2D6 and 2C9 playing only minor metabolic roles (Eiermann et al. 1997; Olesen and Linnet 2001). Rifampin is a "pan-inducer" of 1A2, 3A4, 2C9, and 2C19 (Heimark et al. 1987; Kay et al. 1985; Strayhorn et al. 1997; Wietholtz et al. 1995; Zhou et al. 1990; Zilly et al. 1977). Thus, the addition of rifampin increased the amount of 1A2, 3A4, and 2C9/19 that was available to metabolize the clozapine, resulting in a sharp decline in the clozapine blood level, even though the clozapine dosage had not been changed. There was also a likely contribution from phase II glucuronidative metabolism. Clozapine is a UGT (uridine 5′-diphosphate glucuronosyltransferase) 1A4 substrate (Breyer-Pfaff and Wachsmuth 2001), and rifampin is a UGT1A4 inducer (Ebert et al. 2000). This interaction likely synergized with the extensive P450 interactions mentioned above. Studies have demonstrated that the addition of rifampin to clozapine can produce three- to sixfold decreases in clozapine blood levels (Finch et al. 2002).

References

Breyer-Pfaff U, Wachsmuth H: Tertiary N-glucuronides of clozapine and its metabolite desmethylclozapine in patient urine. Drug Metab Dispos 29(10): 1343–1348, 2001

Ebert U, Thong NQ, Oertel R, et al: Effects of rifampin and cimetidine on pharmacokinetics and pharmacodynamics of lamotrigine in healthy subjects. Eur J Clin Pharmacol 56(4):299–304, 2000

Eiermann B, Engel G, Johansson I, et al: The involvement of CYP1A2 and CYP3A4 in the metabolism of clozapine. Br J Clin Pharmacol 44(5):439–446, 1997

Finch CF, Chrisman CR, Baciewicz AM, et al: Rifampin and rifabutin drug interactions: an update. Arch Intern Med 162(9):985–992, 2002

Heimark LD, Gibaldi M, Trager WF, et al: The mechanism of the warfarin-rifampin drug interaction. Clin Pharmacol Ther 42(4):388–394, 1987

Kay L, Kampmann JP, Svendsen TL, et al: Influence of rifampicin and isoniazid on the kinetics of phenytoin. Br J Clin Pharmacol 20(4):323–326, 1985

Olesen OV, Linnet K: Contributions of five human cytochrome P450 isoforms to the N-demethylation of clozapine in vitro at low and high concentrations. J Clin Pharmacol 41(8):823–832, 2001

Strayhorn VA, Baciewicz AM, Self TH: Update on rifampin drug interactions, III. Arch Intern Med 157(21):2453–2458, 1997

Wietholtz H, Zysset T, Marschall HU, et al : The influence of rifampin treatment on caffeine clearance in healthy man. J Hepatol 22(1):78–81, 1995

Zhou HH, Anthony LB, Wood AJ, et al: Induction of polymorphic 4′ hydroxylation of S-mephenytoin by rifampicin. Br J Clin Pharmacol 30(3):471–475, 1990

Zilly W, Breimer DD, Richter E: Stimulation of drug metabolism by rifampicin in patients with cirrhosis or cholestasis measured by increased hexobarbital and tolbutamide clearance. Eur J Clin Pharmacol 11(4):287–293, 1977

VIP Psychosis (II)

The patient in the previous case had her clozapine (Clozaril) dosage increased to 2,500 mg/day, which seemed like an immense dosage but only produced a blood level of 468 ng/mL. In the course of her treatment, she was transferred to the "best" facility in the country and eventually transitioned to the care of the "best" psychiatrist in her hometown for outpatient care. However, in the midst of all these transitions, there was a breakdown in communications as to how and why she had come to require such an extraordinary dosage of clozapine. She eventually completed the course of rifampin (Rifadin), at which point she was requiring only every-other-week complete blood count (CBC) blood monitoring for neutropenia. Her rifampin was discontinued just after one of her CBCs. In the 2 weeks before her next CBC, she became severely sedated and ataxic. At first, her family ascribed this to her being overly "dramatic," but they became concerned when she fell down several stairs in their home. They contacted the psychiatrist, who advised them to watch her closely and he would order a clozapine blood level in addition to the planned CBC. However, on the way to have the blood drawn, she experienced a grand mal seizure and was taken to the nearest ER, where her clozapine blood level was found to be 1,853 ng/mL.

Discussion

This is an example of reversal of induction.

Again, clozapine's metabolism occurs primarily at 1A2. The enzymes 3A4 and 2C19 are also significant contributors, with 2D6 and 2C9 playing only minor metabolic roles (Eiermann et al. 1997; Olesen and Linnet 2001). Rifampin is a "pan-inducer" of 1A2, 3A4, 2C9, and 2C19 (Heimark et al. 1987; Kay et al. 1985; Strayhorn et al. 1997; Wietholtz et al. 1995; Zhou et al. 1990; Zilly et al. 1977). The clozapine dosage had been increased to 2,500 mg/day to compensate for rifampin's induction of 1A2, 3A4, 2C9, and 2C19 and to maintain a therapeutic blood level in the face of this induction (Finch et al. 2002). However, with the discontinuation of the rifampin, 1A2, 3A4, 2C9, and 2C19 resumed their lower baseline levels of activity. This led to a marked decrease in the efficiency with which these enzymes metabolized the clozapine, producing an increase in the clozapine blood level, culminating in her toxicity symptoms and seizure. Again, the reversal of induction of glucuronidation enzyme 1A4 also contributed to this increase in the clozapine blood level (Breyer-Pfaff and Wachsmuth 2001; Ebert et al. 2000).

Had the lines of communication not been disrupted by the numerous elective transfers of care, the psychiatrist would have tapered the clozapine back to the pre-rifampin baseline dosage (500 mg/day) in the weeks following discontinuation of the rifampin. This would hopefully have avoided any acute increases in the clozapine blood level and accompanying complications.

References

Breyer-Pfaff U, Wachsmuth H: Tertiary N-glucuronides of clozapine and its metabolite desmethylclozapine in patient urine. Drug Metab Dispos 29(10): 1343–1348, 2001

Ebert U, Thong NQ, Oertel R, et al: Effects of rifampin and cimetidine on pharmacokinetics and pharmacodynamics of lamotrigine in healthy subjects. Eur J Clin Pharmacol 56(4):299–304, 2000

Eiermann B, Engel G, Johansson I, et al: The involvement of CYP1A2 and CYP3A4 in the metabolism of clozapine. Br J Clin Pharmacol 44(5):439–446, 1997

Finch CF, Chrisman CR, Baciewicz AM, et al: Rifampin and rifabutin drug interactions: an update. Arch Intern Med 162(9):985–992, 2002

Heimark LD, Gibaldi M, Trager WF, et al: The mechanism of the warfarin-rifampin drug interaction. Clin Pharmacol Ther 42(4):388–394, 1987

Kay L, Kampmann JP, Svendsen TL, et al: Influence of rifampicin and isoniazid on the kinetics of phenytoin. Br J Clin Pharmacol 20(4):323–326, 1985

Olesen OV, Linnet K: Contributions of five human cytochrome P450 isoforms to the N-demethylation of clozapine in vitro at low and high concentrations. J Clin Pharmacol 41(8):823–832, 2001

Strayhorn VA, Baciewicz AM, Self TH: Update on rifampin drug interactions, III. Arch Intern Med 157(21):2453–2458, 1997

Wietholtz H, Zysset T, Marschall HU, et al: The influence of rifampin treatment on caffeine clearance in healthy man. J Hepatol 22(1):78–81, 1995

Zhou HH, Anthony LB, Wood AJ, et al: Induction of polymorphic 4′ hydroxylation of S-mephenytoin by rifampicin. Br J Clin Pharmacol 30(3):471–475, 1990

Zilly W, Breimer DD, Richter E: Stimulation of drug metabolism by rifampicin in patients with cirrhosis or cholestasis measured by increased hexobarbital and tolbutamide clearance. Eur J Clin Pharmacol 11(4):287–293, 1977

Wired

A 19-year-old college student had just been diagnosed with obsessive-compulsive disorder (OCD). He visited the student mental health clinic, and the clinic psychiatrist started him on fluvoxamine (Luvox), which was eventually titrated to a dosage of 200 mg/day, which proved to be both effective and well tolerated. In years past, the patient had tried a few cups of coffee. While he had enjoyed the taste, he had never become a habitual coffee drinker. However, one afternoon he went with some of his friends for lunch at a local coffeehouse and had two cappuccinos. That night, he could not sleep because he felt agitated, jittery, and "wired." He also had to urinate every 90 minutes, and he became quite thirsty. He reported these events to his psychiatrist. After a brief discussion over the phone, the patient decided that he had enjoyed such a positive response from the fluvoxamine that he did not want to change medicines, but that he would steer clear of all caffeinated beverages in the future.

Discussion

This is an example of a substrate added to an inhibitor.

Caffeine is a 1A2 substrate (Miners and Birkett 1996), and fluvoxamine is a strong 1A2 inhibitor (Brosen 1995). The presence of the fluvoxamine impaired the ability of 1A2 to efficiently metabolize caffeine. When a normal amount of caffeine was ingested, the lack of efficient 1A2 metabolism led to a much greater blood level of caffeine, which also lasted much longer in the bloodstream, than would otherwise have been expected. Studies have demonstrated that adding fluvoxamine will prolong the half-life of caffeine more than sixfold (Jeppesen et al. 1996)! So in this case, the presence of fluvoxamine turned an innocent trip to a coffeehouse into a brush with caffeine toxicity.

References

Brosen K: Drug interactions and the cytochrome P450 system: the role of cytochrome P450 1A2. Clin Pharmacokinet 29 (suppl 1):20–25, 1995

Jeppesen U, Loft S, Poulson HE, et al: A fluvoxamine-caffeine interaction study. Pharmacogenetics 6(3):213–222, 1996

Miners JO, Birkett DJ: The use of caffeine as a metabolic probe for human drug metabolizing enzymes. Gen Pharmacol 27(2):245–249, 1996

Shake, Rattle, and Roll

A 37-year-old outpatient with schizoaffective disorder, bipolar type, experienced a positive response to clozapine (Clozaril), 600 mg/day (blood level=502 ng/mL), and lithium (Eskalith), 900 mg/day (blood level=0.8 mEq/L), after several previously unsuccessful medication regimens. However, several repetitive, perseverative ritual behaviors began to emerge (for example, excessive hand washing and hair brushing, and flooding the toilets with toilet paper). The psychiatrist thought it likely that these behaviors were a side effect of the clozapine, but he was loathe to discontinue one of the only medications that had been of benefit for the patient's entrenched psychotic symptoms and impulsivity. He therefore decided to treat these compulsive behaviors by adding fluvoxamine (Luvox), which he titrated to a dosage of 150 mg/day over a 10-day period. After the first 5 days, the patient complained about significant sedation, but the psychiatrist reassured him that this was a commonly encountered side effect with fluvoxamine and that the sedation should be transient. On day 11, however, the patient experienced a grand mal seizure. His family called 911, and an ambulance transported him to the nearest ER. There, the blood level of clozapine was found to be 2,112 ng/mL.

Discussion

This is an example of an inhibitor added to a substrate.

Clozapine's metabolism occurs primarily at 1A2. Enzymes 3A4 and 2C19 are also significant contributors, with 2D6 and 2C9 playing only minor metabolic roles (Eiermann et al. 1997; Olesen and Linnet 2001). Fluvoxamine is a strong inhibitor of 1A2, 2C9, and 2C19 and a moderate inhibitor of 3A4 (Christensen et al. 2002; Niemi et al. 2001; von Moltke et al. 1995). Thus, the addition of the fluvoxamine led to an impairment of clozapine's metabolism by these P450 enzymes, resulting in a more than fourfold increase in the clozapine blood level, even though the clozapine dosage had remained constant throughout (Brosen 1995). After an initial period in which the patient experienced sedation, this toxic clozapine blood level lowered the patient's seizure threshold enough to result in a new-onset grand mal seizure.

References

Brosen K: Drug interactions and the cytochrome P450 system: the role of cytochrome P450 1A2. Clin Pharmacokinet 29 (suppl 1):20–25, 1995

Christensen M, Tybring G, Mihara K, et al: Low daily 10-mg and 20-mg doses of fluvoxamine inhibit the metabolism of both caffeine (cytochrome P4501A2) and omeprazole (cytochrome P4502C19). Clin Pharmacol Ther 71(3):141–152, 2002

Eiermann B, Engel G, Johansson I, et al: The involvement of CYP1A2 and CYP3A4 in the metabolism of clozapine. Br J Clin Pharmacol 44(5):439–446, 1997

Niemi M, Backman JT, Neuvonen M, et al: Effects of fluconazole and fluvoxamine on the pharmacokinetics and pharmacodynamics of glimepiride. Clin Pharmacol Ther 69(4):194–200, 2001

Olesen OV, Linnet K: Contributions of five human cytochrome P450 isoforms to the N-demethylation of clozapine in vitro at low and high concentrations. J Clin Pharmacol 41(8):823–832, 2001

von Moltke LL, Greenblatt DJ, Court MH, et al: Inhibition of alprazolam and desipramine hydroxylation in vitro by paroxetine and fluvoxamine: comparison with other selective serotonin reuptake inhibitor antidepressants. J Clin Psychopharmacol 15(2):125–131, 1995

Smoking Gun (I)

A 25-year-old man with bipolar I disorder had been recently discharged after a 4-week stay at an inpatient psychiatric facility, where his acute manic episode had been successfully treated with olanzapine (Zyprexa) monotherapy, at a dosage of 20 mg/day. He was a habitual two-pack-a-day smoker, and the inpatient facility he had just left had a strict no-smoking policy. He immediately resumed his usual level of smoking after discharge from the hospital. Within 3 weeks, he was readmitted after assaulting a man at a bus station who would not give him bus fare for a trip to Yucca Mountain, Nevada, where he intended to protest the planned deposition of nuclear waste. Noncompliance with his medication was suspected, but he had his supply of medications with him, and the remaining number of olanzapine tablets was exactly the same as would be expected had he been rigorously compliant with his regimen. This patient was notoriously unreliable at following up on the laboratory studies that are necessary to remain on most conventional mood stabilizers, so the psychiatrist again sought to stabilize him with atypical antipsychotics. A trial of quetiapine (Seroquel) was tried, and the patient again improved and was discharged, but this time he remained stable for a sustained period. At the time, the psychiatrist attributed this sequence of events to idiosyncratic differences in medication response in a given patient. It was not until after he attended a P450 grand rounds presentation that he revised this assessment.

Discussion

This is an example of an inducer added to a substrate.

Olanzapine is a 1A2 substrate, although it is also metabolized by 2D6 and phase II glucuronidation (Callaghan et al. 1999). Tobacco use, via cigarette smoking, is a significant 1A2 inducer (Schrenk et al. 1998; Zevin and Benowitz 1999). When the patient resumed smoking, he significantly increased the amount of 1A2 that was available to metabolize the olanzapine, resulting in a decrease in the blood level of olanzapine (Skogh et al. 2002). Studies have demonstrated that smoking may increase the clearance of olanzapine by as much as 40% (Cozza et al. 2003). Although olanzapine blood levels are not typically obtained in standard clinical practice, it seems clear that the increases in olanzapine's metabolism and the subsequent decreases in the blood levels of olanzapine caused by smoking tobacco can produce significant clinical decompensations. The use of tobacco should be one of the issues that are closely followed when prescribing olanzapine (and clozapine).

References

Callaghan JT, Bergstrom RF, Ptak LR, et al: Olanzapine: pharmacokinetic and pharmacodynamic profile. Clin Pharmacokinet 37(3):177–193, 1999

Cozza KL, Armstrong SC, Oesterheld JR: Concise Guide to Drug Interaction Principles for Medical Practice: Cytochrome P450s, UGTs, P-glycoproteins, 2nd Edition. Washington, DC, American Psychiatric Publishing, 2003

Schrenk D, Brockmeier D, Morike K, et al: A distribution study of CYP1A2 phenotypes among smokers and non-smokers in a cohort of healthy Caucasian volunteers. Eur J Clin Pharmacol 53(5):361–367, 1998

Skogh E, Reis M, Dahl ML, et al: Therapeutic drug monitoring data on olanzapine and its N-demethyl metabolite in the naturalistic clinical setting. Ther Drug Monit 24(4):518–526, 2002

Zevin S, Benowitz NL: Drug interactions with tobacco smoking: an update. Clin Pharmacokinet 36(6):425–438, 1999

Smoking Gun (II)

A 47-year-old woman with chronic paranoid schizophrenia was being well maintained on clozapine (Clozaril), 700 mg/day (blood level=455 ng/mL). During a routine checkup with her internist, he explained to her the importance of smoking cessation, especially in view of her moderate hypertension (even while on clozapine) and family history of early demise from cardiac events. With the support and encouragement of her family and friends, she used a nicotine patch (prescribed by her internist) and abruptly stopped smoking. Within the next 3 weeks, however, she developed increasing sedation, dizziness when rising from a sitting position, blurry vision, and constipation. She reported her difficulties to her internist and psychiatrist. Her psychiatrist ordered an immediate clozapine blood level with instructions to then decrease her clozapine dosage to 400 mg/day for the time being. The clozapine blood level was 893 ng/mL. After 4 days, the patient reported a remission of the aforementioned symptoms, and the psychiatrist ordered another clozapine blood level, which was now 485 ng/mL. The psychiatrist conferred with the patient, and the new plan was to dose the clozapine stably at 400 mg/day, but he cautioned the patient that she needed to inform him as soon as possible if she resumed smoking.

Discussion

This is an example of reversal of induction.

Clozapine's metabolism occurs primarily at 1A2. The enzymes 3A4 and 2C19 are also significant contributors, with 2D6 and 2C9 playing only minor metabolic roles (Eiermann et al. 1997; Olesen and Linnet 2001). Tobacco use, via cigarette smoking, is a significant 1A2 inducer (Schrenk et al. 1998; Zevin and Benowitz 1999). The previous clozapine dosage had been determined in the presence of constant tobacco use, which caused a stable increase in the amount of 1A2 that was available to metabolize the clozapine. This led to a rate of metabolism of clozapine that was significantly greater than if the patient had not been smoking. When she stopped smoking, the "extra" 1A2 that had been induced by the tobacco "died off" over the following few weeks, resulting in a return to presmoking (lower) levels of 1A2 activity. Since the metabolism of clozapine by 1A2 was now decreased, this led to an almost doubling of the patient's clozapine blood level, even though the dosage had been constant until then (Zullino et al. 2002). The resulting state of mild clozapine toxicity caused the patient to experience orthostasis, sedation, and anticholinergic symptoms (blurry vision and constipation). The psychiatrist successfully compensated for this reversal of induction by decreasing the clozapine dosage from 700 mg/day to 400 mg/day, at which time these symptoms quickly remitted. However, the psychiatrist prudently cautioned the patient to inform him if she resumed smoking, as the clozapine dosage would again have to be increased in order to prevent a decrease in clozapine blood levels and a subsequent loss of efficacy.

While smoking cessation is a laudable goal, these two cases highlight the need to be prospectively aware of how changes, or briefly enforced differences, in smoking behavior can affect drug metabolism in various ways.

References

Eiermann B, Engel G, Johansson I, et al: The involvement of CYP1A2 and CYP3A4 in the metabolism of clozapine. Br J Clin Pharmacol 44(5):439–446, 1997

Olesen OV, Linnet K: Contributions of five human cytochrome P450 isoforms to the N-demethylation of clozapine in vitro at low and high concentrations. J Clin Pharmacol 41(8):823–832, 2001

Schrenk D, Brockmeier D, Morike K, et al: A distribution study of CYP1A2 phenotypes among smokers and non-smokers in a cohort of healthy Caucasian volunteers. Eur J Clin Pharmacol 53(5):361–367, 1998

Zevin S, Benowitz NL: Drug interactions with tobacco smoking: an update. Clin Pharmacokinet 36(6):425–438, 1999

Zullino DF, Delessert D, Eap CB, et al: Tobacco and cannabis smoking cessation can lead to intoxication with clozapine or olanzapine. Int Clin Psychopharmacol 17(3):141–143, 2002

Java Jitters

A 55-year-old diabetic man noticed that a sore on the underside of his right foot was growing in size and becoming more red and warm. Although he felt no acute discomfort, he was aware that his neuropathy might deaden the pain even if a serious infection was present. He therefore visited his internist, who diagnosed cellulitis and prescribed ciprofloxacin (Cipro). The patient had long relied on two strong cups of coffee to help him "start the day." By day 4 of his ciprofloxacin regimen, he was experiencing frank jitteriness and palpitations. He reported this to his internist, who consulted his affiliated hospital's pharmacist. Following this discussion, the internist instructed the patient to drink decaffeinated coffee and avoid other caffeinated beverages for the duration of the ciprofloxacin treatment course.

Discussion

This is an example of an inhibitor added to a substrate.

Caffeine, in addition to being our most commonly used psychoactive substance, is a 1A2 substrate (Miners and Birkett 1996). Ciprofloxacin is a strong 1A2 inhibitor (Batty et al. 1995). The addition of the ciprofloxacin markedly impaired the ability of 1A2 to efficiently metabolize caffeine (Mizuki et al. 1996). This led to a marked increase (often two- to threefold) in the blood level of caffeine following the patient's usual consumption of two cups of coffee (K.L. Cozza, S.C. Armstrong, personal communication, May 2002), which caused his symptoms of agitation and palpitations.

References

Batty KT, Davis TM, Ilett KF, et al: The effect of ciprofloxacin on theophylline pharmacokinetics in healthy subjects. Br J Clin Pharmacol 39(3):305–311, 1995

Miners JO, Birkett DJ: The use of caffeine as a metabolic probe for human drug metabolizing enzymes. Gen Pharmacol 27(2):245–249, 1996

Mizuki Y, Fujiwara I, Yamaguchi T: Pharmacokinetic interactions related to the chemical structures of fluoroquinolones. J Antimicrob Chemother 37 (suppl A): 41–55, 1996

Anxious About Anthrax

A 27-year-old woman with chronic paranoid schizophrenia had been maintaining a reasonable degree of clinical stability while taking clozapine (Clozaril), 600 mg/day (blood level=410 ng/mL), although some chronic paranoia was generally present. Over the course of 1 week, she developed a persistent cough and low-grade fever, whereupon she became convinced that she had contracted anthrax. She convinced her internist to prescribe a course of ciprofloxacin (Cipro). Within 5 days, she became progressively weaker, more sedated, and dizzy, although her cough did begin to clear. Nonetheless, she was convinced that she had not begun the ciprofloxacin early enough and that these new symptoms were the result of infection with anthrax. She finally called an ambulance to transport her to the nearest ER. The ER physician was able to reassure her that she did not have anthrax and that her weakness, dizziness, and sedation were attributable to her mild clozapine toxicity, as her blood level had risen to 734 ng/mL.

Discussion

This is an example of an inhibitor added to a substrate.

Clozapine's metabolism occurs primarily at 1A2. Enzymes 3A4 and 2C19 are also significant contributors, with 2D6 and 2C9 playing only minor metabolic roles (Eiermann et al. 1997; Olesen and Linnet 2001). Ciprofloxacin is a strong inhibitor of both 1A2 and 3A4 (Batty et al. 1995; McLellan et al. 1996). With the addition of the ciprofloxacin, the ability of 1A2 and 3A4 to significantly contribute to the metabolism of clozapine was markedly impaired. Since these two enzymes handle the majority of clozapine's metabolism, their inhibition led to an almost doubling of the clozapine blood level, even though the clozapine dosage had not been changed (Raaska and Neuvonen 2000). The resulting state of mild clozapine toxicity explained the patient's emerging somatic symptoms.

References

Batty KT, Davis TM, Ilett KF, et al: The effect of ciprofloxacin on theophylline pharmacokinetics in healthy subjects. Br J Clin Pharmacol 39(3):305–311, 1995

Eiermann B, Engel G, Johansson I, et al: The involvement of CYP1A2 and CYP3A4 in the metabolism of clozapine. Br J Clin Pharmacol 44(5):439–446, 1997

McLellan RA, Drobitch RK, Monshouwer M, et al: Fluoroquinolone antibiotics inhibit cytochrome P450–mediated microsomal drug metabolism in rat and human. Drug Metab Dispos 24(10):1134–1138, 1996

Olesen OV, Linnet K: Contributions of five human cytochrome P450 isoforms to the N-demethylation of clozapine in vitro at low and high concentrations. J Clin Pharmacol 41(8):823–832, 2001

Raaska K, Neuvonen PJ: Ciprofloxacin increases serum clozapine and N-desmethylclozapine: a study in patients with schizophrenia. Eur J Clin Pharmacol 56(8):585–589, 2000

Caffeine Complications

A 32-year-old man with schizoaffective disorder, depressive type, and significant associated anxiety had done reasonably well for the past year on clozapine (Clozaril), 300 mg/day (blood level=589 ng/mL); fluoxetine (Prozac), 40 mg/day; and diazepam (Valium), 10 mg/day. He recently started a volunteer job at the local Veterans Affairs hospital, and he increased his intake of caffeine in order to "perk up" so he could do his best. Specifically, he was taking one 200-mg caffeine tablet each morning, and he consumed two large cups of tea each day. However, after 5 days he did not feel "perky." On the contrary, he experienced increased sedation, ataxia, and blurry vision. He reported these symptoms to his psychiatrist, who checked a clozapine blood level and advised the patient to take a few days off from his new job. This proved to be good advice, as the lack of work responsibilities led the patient to curtail his caffeine intake, and his symptoms abated. The clozapine blood level was found to be 1,145 ng/mL. When the patient reported to his psychiatrist that he was feeling better, he took a more careful history and discovered the patient's change in caffeine intake. A follow-up clozapine blood level was 622 ng/mL (Odom-White and deLeon 1996).

Discussion

This is an example of an inhibitor added to a substrate.

Clozapine's metabolism occurs primarily at 1A2. Enzymes 3A4 and 2C19 are also significant contributors, with 2D6 and 2C9 playing only minor metabolic roles (Eiermann et al. 1997; Olesen and Linnet 2001). Caffeine is generally considered to be a substrate of 1A2 and not a significant inhibitor (Miners and Birkett 1996). However, the presence of the fluoxetine likely changed the "P450 landscape" in ways that made caffeine an effective competitive inhibitor. Fluoxetine (in concert with its active metabolite, norfluoxetine) is a strong inhibitor of 2D6 and a moderate inhibitor of 3A4, 2C9, and 2C19 (Greenblatt et al. 1999; Stevens and Wrighton 1993). The long-standing coadministration of clozapine and fluoxetine may partially explain why a relatively low dosage of clozapine (300 mg/day) generated a somewhat higher than expected, but therapeutic, clozapine blood level (589 ng/mL). (See "Enuresis (I)" [p. 133] for a more complete discussion of the interaction of clozapine and fluoxetine.) However, fluoxetine's inhibition of all of the major non-1A2 P450 enzymes involved in clozapine's metabolism rendered clozapine's metabolism even more primarily dependent on the activity and availability of 1A2 than usual. Thus, when an avidly bound 1A2 substrate like caffeine was added to the regimen, it was able to compete with substrate-binding sites on 1A2 to a sufficient degree to act as a functionally significant competitive inhibitor of 1A2's ability to metabolize the clozapine. This led to an increase in the clozapine blood level, even though there had been no increase in the dosage, which caused the patient to experience increased sedation, ataxia, and blurry vision. These factors explain why the addition of a generally stimulating substance (caffeine) led to paradoxical sedation and anergy. With the removal of the competitive inhibitor, caffeine, the patient's clozapine blood level returned to baseline, and his symptoms remitted.

References

Eiermann B, Engel G, Johansson I, et al: The involvement of CYP1A2 and CYP3A4 in the metabolism of clozapine. Br J Clin Pharmacol 44(5):439–446, 1997

Greenblatt DJ, von Moltke LL, Harmatz JS, et al: Human cytochromes and some newer antidepressants: kinetics, metabolism, and drug interactions. J Clin Psychopharmacol 19 (5, suppl 1):23S–35S, 1999

Miners JO, Birkett DJ: The use of caffeine as a metabolic probe for human drug metabolizing enzymes. Gen Pharmacol 27(2):245–249, 1996

Odom-White A, deLeon J: Clozapine levels and caffeine. J Clin Psychiatry 57(4): 175–176, 1996

Olesen OV, Linnet K: Contributions of five human cytochrome P450 isoforms to the N-demethylation of clozapine in vitro at low and high concentrations. J Clin Pharmacol 41(8):823–832, 2001

Stevens JC, Wrighton SA: Interaction of the enantiomers of fluoxetine and norfluoxetine with human liver cytochromes P450. J Pharmacol Exp Ther 266(2): 964–971, 1993

GI Joe (I)

A 27-year-old man with chronic paranoid schizophrenia was being stably maintained on olanzapine (Zyprexa), 25 mg/day. He developed some persistent heartburn, which led him to visit his primary care physician. After a gastrointestinal (GI) workup, he was diagnosed with gastroesophageal reflux disease, and his doctor prescribed cimetidine (Tagamet), 300 mg qid, for cost reasons. After 5 days, the patient experienced increased sedation and constipation. When his psychiatrist was consulted, he decreased the dosage of olanzapine to 15 mg/day, which promptly led to a remission of these symptoms. The patient remained psychiatrically stable throughout and into the future.

Discussion

This is an example of an inhibitor added to a substrate.

Olanzapine is a 1A2 substrate, although it is also metabolized by 2D6 and phase II glucuronidation (Callaghan et al. 1999). Cimetidine is a "pan-inhibitor" that strongly inhibits 1A2, 2D6, and 3A4 (Martinez et al. 1999). With the addition of the cimetidine, 1A2 and 2D6 were impaired in their ability to contribute to olanzapine's metabolism, although the majority of this inhibition likely occurred at 1A2. Thus, the blood level of olanzapine rose, even though there had been no increase in the olanzapine dosage, which led to the emergence of olanzapine side effects (sedation and constipation) (Pies 2002). The psychiatrist compensated for the decreased metabolism of olanzapine by decreasing the dosage from 25 mg/day to 15 mg/day, which led to a remission of these side effects with a preservation of antipsychotic efficacy.

References

Callaghan JT, Bergstrom RF, Ptak LR, et al: Olanzapine: pharmacokinetic and pharmacodynamic profile. Clin Pharmacokinet 37(3):177–193, 1999
Martinez C, Albet C, Agundez JA, et al: Comparative in vitro and in vivo inhibition of cytochrome P450 CYP1A2, CYP2D6, and CYP3A by H2-receptor antagonists. Clin Pharmacol Ther 65(4):369–376, 1999

Pies R: Cytochromes and beyond: drug interactions in psychiatry. Psychiatric Times, May 2002, pp 48–51

GI Joe (II)

The patient in the previous case had been doing well while taking olanzapine (Zyprexa), 15 mg/day, and cimetidine, 300 mg qid, for 2 years. However, his GI discomfort began to worsen, and his primary care physician decided to discontinue his cimetidine and begin omeprazole (Prilosec), 20 mg/day, until a more comprehensive GI workup could be undertaken. After 5 days, the patient reported feeling more alert and almost "activated," but he remained essentially stable. However, within another 3 weeks, he began to experience threatening auditory hallucinations and the paranoid delusion that his parents were trying to draft him into the army. These symptoms worsened to the point that the patient required hospitalization, to which he agreed. An initial attempt was made to retitrate his olanzapine back to 25 mg/day, but his psychotic symptoms responded only partially to this dosage, even after a period of 4 weeks. After consulting with the hospital pharmacist, the psychiatrist resumed dose titration of the olanzapine, and he found that the patient reachieved a complete response at a dosage of 40 mg/day.

Discussion

This is a combined example of reversal of inhibition and an inducer added to a substrate.

Remember from the previous case that olanzapine is a 1A2 substrate, although it is also metabolized by 2D6 and phase II glucuronidation (Callaghan et al. 1999), and cimetidine is a "pan-inhibitor" that strongly inhibits 1A2, 2D6, and 3A4 (Martinez et al. 1999). When the cimetidine was discontinued, 1A2, 2D6, and 3A4 quickly resumed their previous levels of activity and thus were more efficient in metabolizing the olanzapine than they were in the presence of the cimetidine. This led to a decrease in the blood level of the olanzapine, even though the dosage had not been decreased. If the omeprazole had not been added, the olanzapine dosage could have been increased back to 25 mg/day, and the patient would likely have returned to his asymptomatic baseline at this dosage. However, omeprazole is a 1A2 inducer (Nousbaum et al. 1994). With the addition of the omeprazole, more 1A2 was produced and thus available to metabolize the olanzapine. This led to a further decrease in the blood level of olanzapine (DeVane and Nemeroff 2002), thus requiring even more of a dosage increase to provide antipsychotic efficacy, which was eventually achieved at a dosage of 40 mg/day.

References

Callaghan JT, Bergstrom RF, Ptak LR, et al: Olanzapine: pharmacokinetic and pharmacodynamic profile. Clin Pharmacokinet 37(3):177–193, 1999

DeVane CL, Nemeroff CB: 2002 guide to psychotropic drug interactions. Primary Psychiatry 9(3):28–57, 2002

Martinez C, Albet C, Agundez JA, et al: Comparative in vitro and in vivo inhibition of cytochrome P450 CYP1A2, CYP2D6, and CYP3A by H2-receptor antagonists. Clin Pharmacol Ther 65(4):369–376, 1999

Nousbaum JB, Berthou F, Carlhant D, et al: Four-week treatment with omeprazole increases the metabolism of caffeine. Am J Gastroenterol 89(3):371–375, 1994

GI Joe (III): The Return to Baseline

The patient in the previous case was discharged from the hospital after 2 weeks of taking olanzapine (Zyprexa), 40 mg/day. One month later, after further studies, his primary care physician determined that his GI symptoms were caused by an infection with *Helicobacter pylori.* He prescribed metronidazole (Flagyl) and tetracycline, in addition to the omeprazole (Prilosec), for a 14-day period and then discontinued the metronidazole, tetracycline, and omeprazole. Within 3 weeks, the patient again experienced the same sedation and constipation he had encountered when cimetidine had initially been added to his olanzapine (see "GI Joe [I]," p. 117). The olanzapine was decreased back to a dosage of 25 mg/day, and he again reachieved a state of psychotic symptom control without any notable side effects.

Discussion

This is an example of reversal of induction.

Olanzapine is a 1A2 substrate, although it is also metabolized by 2D6 and phase II glucuronidation (Callaghan et al. 1999). Omeprazole is a 1A2 inducer (Nousbaum et al. 1994). With the discontinuation of the omeprazole, the "extra" 1A2 that had been produced, and that had led to more efficient metabolism of olanzapine, "died off" over the next 2–3 weeks. This resulted in a return to baseline (lower) levels of 1A2 and a return to a baseline (lower) level of metabolism of olanzapine (DeVane and Nemeroff 2002). Thus, the 40 mg/day dosage of olanzapine, which had been necessary to compensate for omeprazole's induction of 1A2, began to produce side effects after the omeprazole was discontinued. A return to the original dosage of olanzapine (25 mg/day) led to a remission of side effects and a preservation of antipsychotic efficacy.

References

Callaghan JT, Bergstrom RF, Ptak LR, et al: Olanzapine: pharmacokinetic and pharmacodynamic profile. Clin Pharmacokinet 37(3):177–193, 1999

DeVane CL, Nemeroff CB: 2002 guide to psychotropic drug interactions. Primary Psychiatry 9(3):28–57, 2002

Nousbaum JB, Berthou F, Carlhant D, et al: Four-week treatment with omeprazole increases the metabolism of caffeine. Am J Gastroenterol 89(3):371–375, 1994

Bedstuck

A 25-year-old man with comorbid bipolar I disorder and obsessive-compulsive disorder (OCD) was hospitalized because of a manic episode. Prior to his hospitalization, he had been maintained on divalproex sodium (Depakote), 1,250 mg/day (blood level=87 μg/mL). Soon after the admission, olanzapine (Zyprexa), 20 mg/day, was added, and his mania remitted over the next 10 days. However, he then experienced a recurrence of his OCD symptoms (severe contamination obsessions with cleaning compulsions). His OCD had responded to fluvoxamine (Luvox) in the past, although use of this medication was kept to a minimum because of the concern about inducing a manic episode. However, since he now was taking both divalproex sodium and olanzapine, his psychiatrist felt that the benefits of treating the OCD symptoms with fluvoxamine outweighed the risks. Watching closely for a manic relapse, the psychiatrist carefully added fluvoxamine and titrated it to a dosage of 100 mg/day. Within 5 days, contrary to the psychiatrist's concerns, the patient was not activated but rather became quite sedated. The patient slept 13 hours a day, and he stayed in his bed for most of the time that he was awake, claiming to be too "wiped out" to participate in groups or walks. Although his sedation acutely overwhelmed his OCD symptomatology, the psychiatrist did not consider this an optimal therapeutic response. After consulting the hospital pharmacist, he opted to discontinue the fluvoxamine and instead initiated a trial of citalopram (Celexa). The patient's sedation then remitted, and he experienced a remission of his OCD symptoms without a manic relapse (for the time being).

Discussion

This is an example of an inhibitor added to a substrate.

Olanzapine is a 1A2 substrate, although it is also metabolized by 2D6 and phase II glucuronidation (Callaghan et al. 1999). Fluvoxamine is a strong inhibitor of 1A2, 2C9, and 2C19 and a moderate inhibitor of 3A4 (Christensen et al. 2002; Niemi et al. 2001; von Moltke et al. 1995). Thus, the addition of fluvoxamine impaired the ability of 1A2 to significantly contribute to the metabolism of the olanzapine. Clearly, 2D6 and phase II glucuronidation were not able to handle the increased metabolic "burden," resulting in an increase in the blood level of olanzapine and oversedation, even though the olanzapine dosage had not been changed. This sedation was especially surprising, since the inpatient treatment team was anticipating a possible antidepressant-induced manic switch. Studies in which fluvoxamine was added to olanzapine have demonstrated a more than doubling of olanzapine blood levels as a result (Weigmann et al. 2001).

References

Callaghan JT, Bergstrom RF, Ptak LR, et al: Olanzapine: pharmacokinetic and pharmacodynamic profile. Clin Pharmacokinet 37(3):177–193, 1999

Christensen M, Tybring G, Mihara K, et al: Low daily 10-mg and 20-mg doses of fluvoxamine inhibit the metabolism of both caffeine (cytochrome P4501A2) and omeprazole (cytochrome P4502C19). Clin Pharmacol Ther 71(3):141–152, 2002

Niemi M, Backman JT, Neuvonen M, et al: Effects of fluconazole and fluvoxamine on the pharmacokinetics and pharmacodynamics of glimepiride. Clin Pharmacol Ther 69(4):194–200, 2001

von Moltke LL, Greenblatt DJ, Court MH, et al: Inhibition of alprazolam and desipramine hydroxylation in vitro by paroxetine and fluvoxamine: comparison with other selective serotonin reuptake inhibitor antidepressants. J Clin Psychopharmacol 15(2):125–131, 1995

Weigmann H, Gerek S, Zeisig A, et al: Fluvoxamine but not sertraline inhibits the metabolism of olanzapine: evidence from a therapeutic drug monitoring service. Ther Drug Monit 23(4):410–413, 2001

Assassins

A 42-year-old man with schizoaffective disorder, bipolar type, was admitted to a psychiatric unit after 3 months of medication noncompliance.

On admission, he endorsed the strong paranoid delusion that his ex-girlfriend's "snipers" were tracking him with the objective of assassinating him the moment he let his guard down. In his efforts to defend himself from this perceived threat, he had been extremely violent both in the community and in the ER. He was started on his usual 20 mg/day dosage of olanzapine (Zyprexa), and then 3 days later carbamazepine (Tegretol) was added and titrated to a dosage of 800 mg/day (blood level=8.5 μg/mL). Within 1 week of admission, he began to experience a significant decrease in the intensity of his delusions, and he was discharged after 10 days in the hospital. However, 1 week after discharge, he again voiced concerns about being "stalked" and "hunted down like a dog" by "those assassins." After speaking with his psychiatrist, he consented to a voluntary readmission to the hospital. During the admission process, the patient's sister stated that she had been dispensing the patient's medications to him, and she could verify that he had been completely compliant with his prescribed medications.

Discussion

This is an example of an inducer added to a substrate.

Olanzapine is a 1A2 substrate, although it is also metabolized by 2D6 and phase II glucuronidation at the 1A4 enzyme (Callaghan et al. 1999; Linnet 2002). Carbamazepine is an inducer of 3A4, 1A2, and glucuronidation enzyme 1A4 (Ketter et al. 1999; Parker et al. 1998; Rambeck et al. 1996; Spina et al. 1996). Thus, when the carbamazepine was added, more P450 1A2 and glucuronidation enzyme 1A4 were produced and therefore available to more efficiently metabolize the olanzapine. Over the course of 2 weeks, this led to a decrease in the blood level of olanzapine, which caused the patient's partial psychotic relapse, even though he had been medication compliant. Studies in which carbamazepine was added to olanzapine have demonstrated a 40% decrease in olanzapine blood levels (Linnet and Olesen 2002). In this case, that would be comparable to the patient's olanzapine dosage being decreased from the effective dosage of 20 mg/day to a dosage of 12 mg/day.

References

Callaghan JT, Bergstrom RF, Ptak LR, et al: Olanzapine: pharmacokinetic and pharmacodynamic profile. Clin Pharmacokinet 37:177–193, 1999

Ketter TA, Frye MA, Cora-Locatelli G, et al: Metabolism and excretion of mood stabilizers and new anticonvulsants. Cell Mol Neurobiol 19:511–532, 1999

Linnet K: Glucuronidation of olanzapine by cDNA-expressed human UDP-glucuronosyltransferases and human liver microsomes. Hum Psychopharmacol 17(5):233–238, 2002

Linnet K, Olesen OV: Free and glucuronidated olanzapine serum concentrations in psychiatric patients: influence of carbamazepine comedication. Ther Drug Monit 24(4):512–517, 2002

Parker AC, Pritchard P, Preston T, et al: Induction of CYP1A2 activity by carbamazepine in children using the caffeine breath test. Br J Clin Pharmacol 45(2):176–178, 1998

Rambeck B, Specht U, Wolf P: Pharmacokinetic interactions of the new antiepileptic drugs. Clin Pharmacokinet 31(4):309–24, 1996

Spina E, Pisani F, Perucca E: Clinically significant pharmacokinetic drug interactions with carbamazepine: an update. Clin Pharmacokinet 31(3):198–214, 1996

Atypical Parkinsonism

A 31-year-old woman with chronic paranoid schizophrenia was successfully being treated with olanzapine (Zyprexa), 20 mg/day. She began to develop symptoms of dysuria, flank pain, and fevers, which led her to visit her primary care physician. He performed appropriate tests and diagnosed acute pyelonephritis. While awaiting urine culture results, he started the patient on ciprofloxacin, 500 mg bid for 7 days. By day 4 on the ciprofloxacin, the patient reported increased sedation and new-onset bradykinesia, stiffness, and a postural tremor in her hands, as well as blurry vision and constipation. The psychiatrist advised her to decrease her olanzapine dosage to 10 mg/day for the remainder of her ciprofloxacin treatment course, and then to resume the drug at a dosage of 20 mg/day. Her parkinsonian symptoms rapidly abated, as did her sedation, blurry vision, and constipation, and she experienced no further difficulties.

Discussion

This is an example of an inhibitor added to a substrate.

Olanzapine is a 1A2 substrate, although it is also metabolized by 2D6 and phase II glucuronidation (Callaghan et al. 1999). Ciprofloxacin is a potent inhibitor of 1A2 (Batty et al. 1995). Thus, the addition of the ciprofloxacin significantly impaired the ability of 1A2 to efficiently contribute to the metabolism of the olanzapine. In this case, this led to a significant increase in the blood level of olanzapine, even though the olanzapine dosage had not been increased (DeVane and Nemeroff 2002; Livesey

2000). Although it is unusual for olanzapine to produce parkinsonian symptoms, this can occur at quite high dosages of olanzapine in a dose-dependent manner. After ciprofloxacin was added, the patient's 20 mg/day of olanzapine generated a blood level that was high enough to produce extrapyramidal symptoms. Also, some anticholinergic symptoms (blurry vision and constipation) appeared to be due to this increased blood level of olanzapine (Livesey 2000).

References

Batty KT, Davis TM, Ilett KF, et al: The effect of ciprofloxacin on theophylline pharmacokinetics in healthy subjects. Br J Clin Pharmacol 39(3):305–311, 1995

Callaghan JT, Bergstrom RF, Ptak LR, et al: Olanzapine: pharmacokinetic and pharmacodynamic profile. Clin Pharmacokinet 37(3):177–193, 1999

DeVane CL, Nemeroff CB: 2002 guide to psychotropic drug interactions. Primary Psychiatry 9(3):28–57, 2002

Livesey J: Psychotropic drug interactions. September 2000. Available at: http://www.nzhpa.org.nz/Psych_drugint.pdf. Accessed September 6, 2002

Unsuspected Synergy

A 42-year-old woman with a history of recurrent psychotic depression was being stably maintained on fluvoxamine (Luvox), 200 mg/day, and olanzapine (Zyprexa), 10 mg/day. Prior to this period of stability, she had not tolerated trials with fluoxetine (Prozac) or sertraline (Zoloft) because of headaches, night sweats, and other such side effects. With achievement of a greater level of stability than had been achieved previously, she and her psychiatrist were addressing more quality-of-life issues. The patient reported that her libido had been virtually nonexistent while she had been taking all of these selective serotonin reuptake inhibitors (SSRIs), including her current fluvoxamine. The patient and psychiatrist agreed to a trial of a different antidepressant. Since the loss of libido appeared to be an SSRI "class effect" for this patient, they opted for a trial of mirtazapine (Remeron). A gradual crossover titration, with mirtazapine substituting for fluvoxamine, was completed. There were no side effects as a result of this crossover, but within 1 month of this change, the patient began to hear voices telling her she was a "bad wife and mother." Her mood began to decline, and she experienced insomnia and heightened anxiety. Her psychiatrist increased the mirtazapine dosage from 30 mg/day to 45 mg/day and also increased the dosage of olanzapine from 10 mg/day to 20 mg/day. Within 2 weeks of these medication changes, the patient was no longer experiencing auditory hallucinations and her mood began to improve. She remained on this regimen, and no side effects emerged over time.

Discussion

This is an example of reversal of inhibition.

Olanzapine is a 1A2 substrate, although it is also metabolized by 2D6 and phase II glucuronidation (Callaghan et al. 1999). Fluvoxamine is a strong inhibitor of 1A2, 2C9, and 2C19 and a moderate inhibitor of 3A4 (Christensen et al. 2002; Niemi et al. 2001; von Moltke et al. 1995). Thus, the presence of the fluvoxamine impaired the ability of 1A2 to significantly contribute to the metabolism of the olanzapine, yielding a higher blood level of olanzapine than would have been expected had fluvoxamine not been present. Studies in which fluvoxamine was added to olanzapine have demonstrated a more than doubling of olanzapine blood levels as a result (Weigmann et al. 2001). Therefore, with the discontinuation of fluvoxamine, 1A2 was able to resume its baseline (higher) level of metabolic activity, leading to a decrease (probably by about 50%) in the blood level of olanzapine, even though the olanzapine dosage had not been decreased. (Mirtazapine has no clinically relevant P450 or glucuronidative inhibitory effects [Remeron (package insert) 2002].) This decrease in the olanzapine blood level caused the patient to again experience psychotic symptoms and an emerging recurrence of depression. When the psychiatrist increased the olanzapine dosage to 20 mg/day, he was likely producing approximately the same olanzapine blood level that had been achieved when the patient was taking 10 mg/day in addition to the fluvoxamine; hence, no emerging side effects would be expected, even in the face of this increase in olanzapine dosage.

References

Callaghan JT, Bergstrom RF, Ptak LR, et al: Olanzapine: pharmacokinetic and pharmacodynamic profile. Clin Pharmacokinet 37(3):177–193, 1999

Christensen M, Tybring G, Mihara K, et al: Low daily 10-mg and 20-mg doses of fluvoxamine inhibit the metabolism of both caffeine (cytochrome P4501A2) and omeprazole (cytochrome P4502C19). Clin Pharmacol Ther 71(3):141–152, 2002

Niemi M, Backman JT, Neuvonen M, et al: Effects of fluconazole and fluvoxamine on the pharmacokinetics and pharmacodynamics of glimepiride. Clin Pharmacol Ther 69(4):194–200, 2001

Remeron (package insert). West Orange, NJ, Organon Inc, May 2002

von Moltke LL, Greenblatt DJ, Court MH, et al: Inhibition of alprazolam and desipramine hydroxylation in vitro by paroxetine and fluvoxamine: comparison with other selective serotonin reuptake inhibitor antidepressants. J Clin Psychopharmacol 15(2):125–131, 1995

Weigmann H, Gerek S, Zeisig A, et al: Fluvoxamine but not sertraline inhibits the metabolism of olanzapine: evidence from a therapeutic drug monitoring service. Ther Drug Monit 23(4):410–413, 2001

Conspiracy Theory

A 21-year-old man with bipolar I disorder was being maintained on olanzapine (Zyprexa), 20 mg/day. He had been compliant with this medication, but during finals week at his college he became acutely manic and required psychiatric hospitalization. He had not tolerated previous trials of lithium and divalproex sodium (Depakote) because of tremors on lithium and hair loss on divalproex (not prevented by vitamins with zinc and selenium), so he and his psychiatrist agreed on a trial of carbamazepine (Tegretol), titrated to a dosage of 1,000 mg/day (blood level=9.6 μg/mL). He experienced improvement in his manic symptoms by day 7 while taking 1,000 mg/day of carbamazepine. He was then discharged from the hospital. However, 10 days later, he was again grandiose and paranoid, believing that his professors were "in cahoots with the Masons" to secretly control the leadership of all the G7 nations and that only he could foil this plot. After sharing his concerns with his psychiatrist, he did consent to rehospitalization, to formulate an appropriate defense strategy from a safe location. After consulting with the hospital pharmacist, the psychiatrist titrated the olanzapine dosage to 40 mg/day, which eventually produced a remission of psychosis and no side effects.

Discussion

This is an example of an inducer added to a substrate.

Olanzapine is a 1A2 substrate, although it is also metabolized by 2D6 and phase II glucuronidation at the 1A4 enzyme (Callaghan et al. 1999; Linnet 2002). Carbamazepine is an inducer of 3A4, 1A2, and glucuronidation enzyme 1A4 (Ketter et al. 1999; Parker et al. 1998; Rambeck et al. 1996; Spina et al. 1996). Thus, over the course of 2–3 weeks, the addition of carbamazepine led to an increase in the amounts of all these enzymes, with the result that the olanzapine was more quickly and efficiently metabolized, causing a corresponding decline in the olanzapine blood level.

This decrease in the olanzapine blood level led to the reemergence of paranoid and grandiose delusions.

Studies in which carbamazepine was added to olanzapine have found significant (roughly 40%) decreases in olanzapine blood levels (Linnet and Olesen 2002). On the basis of this information, the hospital pharmacist advised the psychiatrist to double the olanzapine dosage to 40 mg/day, thus compensating for carbamazepine's inductive effects.

References

Callaghan JT, Bergstrom RF, Ptak LR, et al: Olanzapine: pharmacokinetic and pharmacodynamic profile. Clin Pharmacokinet 37(3):177–193, 1999

Ketter TA, Frye MA, Cora-Locatelli G, et al: Metabolism and excretion of mood stabilizers and new anticonvulsants. Cell Mol Neurobiol 19(4):511–532, 1999

Linnet K: Glucuronidation of olanzapine by cDNA-expressed human UDP-glucuronosyltransferases and human liver microsomes. Hum Psychopharmacol 17(5):233–238, 2002

Linnet K, Olesen OV: Free and glucuronidated olanzapine serum concentrations in psychiatric patients: influence of carbamazepine comedication. Ther Drug Monit 24(4):512–517, 2002

Parker AC, Pritchard P, Preston T, et al: Induction of CYP1A2 activity by carbamazepine in children using the caffeine breath test. Br J Clin Pharmacol 45(2): 176–178, 1998

Rambeck B, Specht U, Wolf P: Pharmacokinetic interactions of the new antiepileptic drugs. Clin Pharmacokinet 31(4):309–24, 1996

Spina E, Pisani F, Perucca E: Clinically significant pharmacokinetic drug interactions with carbamazepine: an update. Clin Pharmacokinet 31(3):198–214, 1996

Playing With Fire

A 32-year-old man with schizoaffective disorder, bipolar type, was hospitalized following a failed trial of ziprasidone (Geodon), 160 mg/day, and divalproex sodium (Depakote), 1,500 mg/day (blood level=115 μg/mL). The patient was admitted in a flagrantly paranoid state. He had become acutely threatening to neighbors who he believed were persecuting him, thus necessitating the admission. He had already failed trials with every atypical antipsychotic except for clozapine (Clozaril). Previous "mood stabilizers" that had been tried included lithium, gabapentin (Neurontin), and the current divalproex. His psychiatrist therefore decided to initiate a clozapine trial. He titrated the drug up to a dosage of 500 mg/day, which yielded a clozapine blood level of 453 ng/mL. The patient was less frankly delusional and he denied further concerns about his neighbors, although he still displayed remarkably poor frustration tolerance. When frustrated, he would very quickly revert back to a paranoid stance and become acutely threatening toward staff and peers alike, thus requiring locked-door seclusion on several occasions.

Because of the patient's volatility and lability, his psychiatrist decided to begin a trial of carbamazepine (Tegretol), which the patient had never previously tried. He titrated the dosage of carbamazepine up to 1,000 mg/day over the course of 2 weeks (blood level=8.7 μg/mL). In the first week after this carbamazepine dosage had been reached, the patient displayed less agitation and more of an ability to cope with small frustrations. However, 1 week later, the patient's paranoia rapidly grew more intense, and he accordingly became more threatening again. Before embarking on any major course of action, the psychiatrist ordered another set of blood levels for clozapine and carbamazepine. While the carbamazepine blood level had not changed appreciably, the clozapine blood level was now 211 ng/mL. The psychiatrist consulted the hospital pharmacist, who strongly advised him to discontinue the carbamazepine, rather than to try to raise the clozapine dosage, and to use sedative agents for short-term behavioral control.

Discussion

This is an example of an inducer added to a substrate.

Clozapine's metabolism occurs primarily at 1A2. Enzymes 3A4 and 2C19 are also significant contributors, with 2D6 and 2C9 playing only minor metabolic roles (Eiermann et al. 1997; Olesen and Linnet 2001). Clozapine is also metabolized in part by the phase II glucuronidation enzyme 1A4 (Breyer-Pfaff and Wachsmuth 2001). Carbamazepine is an inducer of 3A4, 1A2, and the glucuronidation enzyme 1A4 (Ketter et al. 1999; Parker et al. 1998; Rambeck et al. 1996; Spina et al. 1996). Thus, the addition of the carbamazepine led to an increased production of 1A2, 3A4, and the phase II glucuronidation enzyme 1A4. These various enzymes were therefore able to more efficiently metabolize the clozapine (DeVane and Nemeroff 2002), with the result that the clozapine blood level dropped by about 50% (typical for this scenario), even though the dosage of clozapine had not been decreased. This drop in the clozapine blood level caused the patient to experience a resurgence of his paranoid delusions and threatening behavior. The discontinuation of the carbamazepine would be expected to yield a return to the baseline clozapine blood level in 2–3 weeks.

Apart from this complex pharmacokinetic interaction, there is a potential pharmacodynamic interaction of great importance and concern. Both of these agents are capable of significant bone marrow suppression, with clozapine causing a roughly 1% incidence of agranulocytosis and carbamazepine potentially causing aplastic anemia and other blood dyscrasias. The marrow suppression of these agents is certainly additive, and possibly synergistic. Since this medication combination carries with it an increased risk of causing a severe, adverse hematologic event, the clinician should have a compelling reason that would justify exposing the patient to such risks. It was this set of concerns that led the pharmacist to advise discontinuation of the carbamazepine.

References

Breyer-Pfaff U, Wachsmuth H: Tertiary N-glucuronides of clozapine and its metabolite desmethylclozapine in patient urine. Drug Metab Dispos 29(10): 1343–1348, 2001

DeVane CL, Nemeroff CB: 2002 guide to psychotropic drug interactions. Primary Psychiatry 9(3):28–57, 2002

Eiermann B, Engel G, Johansson I, et al: The involvement of CYP1A2 and CYP3A4 in the metabolism of clozapine. Br J Clin Pharmacol 44(5):439–446, 1997

Ketter TA, Frye MA, Cora-Locatelli G, et al: Metabolism and excretion of mood stabilizers and new anticonvulsants. Cell Mol Neurobiol 19(4):511–532, 1999

Olesen OV, Linnet K: Contributions of five human cytochrome P450 isoforms to the N-demethylation of clozapine in vitro at low and high concentrations. J Clin Pharmacol 41(8):823–832, 2001

Parker AC, Pritchard P, Preston T, et al: Induction of CYP1A2 activity by carbamazepine in children using the caffeine breath test. Br J Clin Pharmacol 45(2):176–178, 1998

Rambeck B, Specht U, Wolf P: Pharmacokinetic interactions of the new antiepileptic drugs. Clin Pharmacokinet 31(4):309–324, 1996

Spina E, Pisani F, Perucca E: Clinically significant pharmacokinetic drug interactions with carbamazepine: an update. Clin Pharmacokinet 31(3):198–214, 1996

The Law of Unintended Consequences

A 31-year-old man was initially treated for his presenting complaints of obsessional thinking, dysphoria, and associated depressive symptoms with fluvoxamine (Luvox), 200 mg/day. When his bipolar diathesis (in the form of emerging manic symptoms) became more evident later in the treatment, olanzapine (Zyprexa), 10 mg/day, was added, and this successfully restored a state of euthymia. However, 6 months later, he again began to experience the emergence of hypomanic symptoms. Rather than increase the dosage of olanzapine or add another mood stabilizer, the psychiatrist and patient agreed that the best course would be to discontinue the fluvoxamine. Contrary to expectations, however, full manic symptoms rapidly emerged following the discontinuation of the fluvoxamine. The psychiatrist felt compelled to both increase the olanzapine to 20 mg/day and add divalproex sodium (Depakote), which he titrated to a dosage of 1,500 mg/day (blood level=118 μg/mL). The patient's manic symptoms were eventually stabilized, but he complained about oversedation and feeling "blunted."

Discussion

This is an example of reversal of inhibition.

Olanzapine is a 1A2 substrate, although it is also metabolized by 2D6 and phase II glucuronidation (Callaghan et al. 1999). Fluvoxamine is a strong inhibitor of 1A2, 2C9, and 2C19 and a moderate inhibitor of 3A4 (Christensen et al. 2002; Niemi et al. 2001; von Moltke et al. 1995). Thus, the presence of the fluvoxamine impaired the ability of 1A2 to significantly contribute to the metabolism of the olanzapine, yielding a higher blood level of olanzapine than would have been expected had fluvoxamine not been present. Studies in which fluvoxamine was added to olanzapine have demonstrated a more than doubling of olanzapine blood levels as a result (Weigmann et al. 2001). Therefore, with the discontinuation of fluvoxamine, 1A2 was able to resume its baseline (higher) level of metabolic activity, leading to a decrease (probably by about 50%) in the blood level of olanzapine, even though the olanzapine dosage had not been decreased. This occult decrease in the olanzapine blood level unexpectedly precipitated the development of full manic symptoms, which the psychiatrist addressed by both increasing the olanzapine dosage and adding divalproex. Had this interaction been anticipated and had the olanzapine dosage been increased to 20 mg/day with the discontinuation of the fluvoxamine, divalproex might not have been a necessary addition to the regimen, or a lower divalproex dose and blood level might have produced antimanic efficacy without sedation and anergy.

There are clearly similarities with the case "Unsuspected Synergy" (see p. 124), in that both cases demonstrate a reversal of fluvoxamine's inhibition of olanzapine, although the scenarios are different in terms of patient diagnosis and outcome. In some ways, there is more similarity with "Bedstuck" (see p. 120), which also demonstrates fluvoxamine's inhibition of olanzapine's metabolism, in that the interactions between fluvoxamine and olanzapine were unsuspected and discontinuation of the fluvoxamine therefore produced unexpected results (unanticipated sedation in "Bedstuck" and unanticipated precipitation of mania in this case).

References

Callaghan JT, Bergstrom RF, Ptak LR, et al: Olanzapine: pharmacokinetic and pharmacodynamic profile. Clin Pharmacokinet 37(3):177–193, 1999

Christensen M, Tybring G, Mihara K, et al: Low daily 10-mg and 20-mg doses of fluvoxamine inhibit the metabolism of both caffeine (cytochrome P4501A2) and omeprazole (cytochrome P4502C19). Clin Pharmacol Ther 71(3):141–152, 2002

Niemi M, Backman JT, Neuvonen M, et al: Effects of fluconazole and fluvoxamine on the pharmacokinetics and pharmacodynamics of glimepiride. Clin Pharmacol Ther 69(4):194–200, 2001

von Moltke LL, Greenblatt DJ, Court MH, et al: Inhibition of alprazolam and desipramine hydroxylation in vitro by paroxetine and fluvoxamine: comparison with other selective serotonin reuptake inhibitor antidepressants. J Clin Psychopharmacol 15(2):125–131, 1995

Weigmann H, Gerek S, Zeisig A, et al: Fluvoxamine but not sertraline inhibits the metabolism of olanzapine: evidence from a therapeutic drug monitoring service. Ther Drug Monit 23(4):410–413, 2001

Smoking Gun (III)

A 52-year-old man with multiple musculoskeletal complaints was being chronically maintained on cyclobenzaprine (Flexeril), 15 mg tid. He decided to adopt a healthier lifestyle, which included cessation of smoking. He went "cold turkey" and abruptly stopped smoking. After a difficult 2–3 days, he felt better and proceeded with his new exercise and diet programs. However, over the next 2 weeks, he felt increasingly lethargic and fatigued. He initially attributed this to his body "getting used to exercise," but the feeling of being sedated did not remit even when he abstained from exercise for 3 days. He eventually contacted his internist, who asked about any recent life changes. When the internist learned about the patient's new fitness program and that he had recently stopped smoking, he told the patient to stop taking the cyclobenzaprine for 1 day, then resume at a dose of 10 mg tid. The patient followed his instructions, and he reported no further difficulties after that.

Discussion

This is an example of reversal of induction.

Cyclobenzaprine is a substrate of 1A2 and 3A4 (Wang et al. 1996). Smoking significantly induces the production of 1A2 (Schrenk et al. 1998; Zevin and Benowitz 1999), thus increasing the efficiency with which cyclobenzaprine is metabolized by 1A2 and lowering cyclobenzaprine blood levels. While the patient was smoking, the doses required to maintain a therapeutic blood level of cyclobenzaprine were therefore higher than those needed had he not been smoking. When he stopped smoking, 1A2 returned to its lower baseline level of activity. This led to an increase in the cyclobenzaprine blood level, even though the dosage had not been increased (K.L. Cozza, S.C. Armstrong, personal communication, May 2002), with resulting sedation and lethargy. His internist compensated for this reversal of 1A2's induction by decreasing the dosage of cyclobenzaprine, leading to a remission of these symptoms.

References

Schrenk D, Brockmeier D, Morike K, et al: A distribution study of CYP1A2 phenotypes among smokers and non-smokers in a cohort of healthy Caucasian volunteers. Eur J Clin Pharmacol 53(5):361–367, 1998

Wang RW, Liu L, Cheng H: Identification of human liver cytochrome P450 isoforms involved in the in vitro metabolism of cyclobenzaprine. Drug Metab Dispos 24(7):786–791, 1996

Zevin S, Benowitz NL: Drug interactions with tobacco smoking: an update. Clin Pharmacokinet 36(6):425–438, 1999

New and Improved

A 37-year-old woman with chronic paranoid schizophrenia was being maintained on clozapine (Clozaril), 300 mg/day (blood level=391 ng/mL). The patient was also receiving cimetidine (Tagamet), 300 mg qid, from her primary care physician for persistent indigestion. During a routine visit to her primary care physician, he discontinued her cimetidine and started her on pantoprazole (Protonix), stating that this was a "new and improved" version of the cimetidine. Within 3 weeks, she was experiencing an increase in her paranoid and somatic delusions and auditory hallucinations telling her that she was the daughter of Satan. She consented to a hospital admission. A clozapine blood level in the hospital was only 136 ng/mL.

Discussion

This is an example of reversal of inhibition.

Clozapine's metabolism occurs primarily at 1A2. Enzymes 3A4 and 2C19 are also significant contributors, with 2D6 and 2C9 playing only minor metabolic roles (Eiermann et al. 1997; Olesen and Linnet 2001). Cimetidine is a pan-inhibitor of 1A2, 3A4, 2D6, 2C9, and 2C19 (Martinez et al. 1999; Nation et al. 1990). Thus, the presence of cimetidine impaired the efficiency of all the P450 enzymes that play any significant role in the metabolism of clozapine, and led to a higher blood level of clozapine than would have occurred had the cimetidine not been present (Szymanski et al. 1991). Thus, with the discontinuation of the cimetidine, all these enzymes resumed their higher baseline levels of activity, which led to more efficient metabolism of the clozapine and a significant decrease in the clozapine blood level, even though the dosage had not changed throughout. This decreased clozapine level led to a recurrence of psychotic symptoms. Once the patient was in the hospital, this problem was quickly identified and remedied by increasing the clozapine dosage to 600 mg/day. Her subsequent blood level was 425 ng/mL.

References

Eiermann B, Engel G, Johansson I, et al: The involvement of CYP1A2 and CYP3A4 in the metabolism of clozapine. Br J Clin Pharmacol 44(5):439–446, 1997

Martinez C, Albet C, Agundez JA, et al: Comparative in vitro and in vivo inhibition of cytochrome P450 CYP1A2, CYP2D6, and CYP3A by H2-receptor antagonists. Clin Pharmacol Ther 65(4):369–376, 1999

Nation RL, Evans AM, Milne RW: Pharmacokinetic drug interactions with phenytoin (Part I). Clin Pharmacokinet 18(1):37–60, 1990

Olesen OV, Linnet K: Contributions of five human cytochrome P450 isoforms to the N-demethylation of clozapine in vitro at low and high concentrations. J Clin Pharmacol 41(8):823–832, 2001

Szymanski S, Lieberman JA, Picou D, et al: A case-report of cimetidine-induced clozapine toxicity. J Clin Psychiatry 52(1):21–22, 1991

Enuresis (I)

A 38-year-old woman with schizoaffective disorder, depressive type, had maintained clinical stability with clozapine (Clozaril), 600 mg qhs (blood level=516 ng/mL). She typically experienced a seasonal variation in her mood, and one winter she became more depressed than usual, with especially prominent anergia. After meeting with the patient and discussing treatment options, her psychiatrist decided to prescribe fluoxetine (Prozac), 20 mg/day. After 4 weeks, the patient's depressive symptoms were improved, but she felt even more lethargic and anergic than before, despite sleeping at least 12 hours each night. Her sedation became so profound that she was no longer able to reliably wake up when her bladder was full, leading to progressive enuresis. After several episodes of enuresis, she reported this embarrassing symptom to her psychiatrist. He ordered a clozapine blood level, which had increased to 1,079 ng/mL. He instructed the patient to skip her next dose and then resume the clozapine at a dose of 300 mg hs. Within 1 week, her oversedation and accompanying enuresis remitted, and her next clozapine blood level was 581 ng/mL.

Discussion

This is an example of an inhibitor added to a substrate.

Clozapine's metabolism occurs primarily at 1A2. Enzymes 3A4 and 2C19 are also significant contributors, with 2D6 and 2C9 playing only minor metabolic roles (Eiermann et al. 1997; Olesen and Linnet 2001). Fluoxetine (in concert with its active metabolite, norfluoxetine) is a strong 2D6 inhibitor and a moderate inhibitor of 3A4, 2C9, and 2C19 (Greenblatt et al. 1999; Stevens and Wrighton 1993). The addition of fluoxetine thus impaired the ability of 3A4, 2C9, 2C19, and 2D6 to make significant contributions to the overall metabolism of clozapine. Even though the functioning of 1A2 was basically spared by fluoxetine, this was apparently not sufficient to offset the inhibition of the other enzymes and prevent an increase in the clozapine blood level. In this case, the increase in the clozapine blood level led to increased sedation and enuresis. Studies have found that adding fluoxetine to clozapine is likely to increase clozapine blood levels by just over 50% (Spina et al. 1998).

References

Eiermann B, Engel G, Johansson I, et al: The involvement of CYP1A2 and CYP3A4 in the metabolism of clozapine. Br J Clin Pharmacol 44(5):439–446, 1997

Greenblatt DJ, von Moltke LL, Harmatz JS, et al: Human cytochromes and some newer antidepressants: kinetics, metabolism, and drug interactions. J Clin Psychopharmacol 19 (5, suppl 1):23S–35S, 1999

Olesen OV, Linnet K: Contributions of five human cytochrome P450 isoforms to the N-demethylation of clozapine in vitro at low and high concentrations. J Clin Pharmacol 41(8):823–832, 2001

Spina E, Avenoso A, Facciola G, et al: Effect of fluoxetine on the plasma concentrations of clozapine and its major metabolites in patients with schizophrenia. Int Clin Psychopharmacol 13(3):141–145, 1998

Stevens JC, Wrighton SA: Interaction of the enantiomers of fluoxetine and norfluoxetine with human liver cytochromes P450. J Pharmacol Exp Ther 266(2): 964–971, 1993

Chapter 5

2C9/2C19/2E1/2B6 and Complex Phenytoin Case Vignettes

The remaining major P450 enzymes, 2C9, 2C19, 2E1, and 2B6, account for substantially fewer psychotropic drug interactions than do 2D6, 3A4, and 1A2. Major issues include the following:

- Phenytoin is both a substrate (Cadle et al. 1994; Mamiya et al. 1998) and an inducer (Chetty et al. 1998) of 2C9 and 2C19.
- Fluoxetine and fluvoxamine are inhibitors of 2C9 and 2C19 (Christensen et al. 2002; Greenblatt et al. 1999; Harvey and Preskorn 2001; Niemi et al. 2001; Rasmussen et al. 1998).
- *S*-Warfarin is a 2C9 substrate (Heimark et al. 1987; Linder and Valdes 1999).

There are polymorphisms for 2C9 and 2C19, but they are only infrequently clinically relevant for psychiatrists (Alam and Sharma 2001).

References

Alam DA, Sharma RP: Cytochrome enzyme genotype and the prediction of therapeutic response to psychotropics. Psychiatric Annals 31(12):715–722, 2001

Cadle RM, Zenon GJ 3rd, Rodriguez-Barradas MC, et al: Fluconazole-induced symptomatic phenytoin toxicity. Ann Pharmacother 28(2):191–195, 1994

Chetty M, Miller R, Seymour MA: Phenytoin autoinduction. Ther Drug Monit 20(1):60–62, 1998

Christensen M, Tybring G, Mihara K, et al: Low daily 10-mg and 20-mg doses of fluvoxamine inhibit the metabolism of both caffeine (cytochrome P4501A2) and omeprazole (cytochrome P4502C19). Clin Pharmacol Ther 71(3):141–152, 2002

Greenblatt DJ, von Moltke LL, Harmatz JS, et al: Human cytochromes and some newer antidepressants: kinetics, metabolism, and drug interactions. J Clin Psychopharmacol 19 (5, suppl 1):23S–35S, 1999

Harvey AT, Preskorn SH: Fluoxetine pharmacokinetics and effect on CYP2C19 in young and elderly volunteers. J Clin Psychopharmacol 21(2):161–166, 2001

Heimark LD, Gibaldi M, Trager WF, et al: The mechanism of the warfarin-rifampin drug interaction. Clin Pharmacol Ther 42(4):388–394, 1987

Linder MW, Valdes R Jr: Pharmacogenetics in the practice of laboratory medicine. Mol Diagn 4(4):365–379, 1999

Mamiya K, Ieiri I, Shimamoto J, et al: The effects of genetic polymorphisms of CYP2C9 and CYP2C19 on phenytoin metabolism in Japanese adult patients with epilepsy: studies in stereoselective hydroxylation and population pharmacokinetics. Epilepsia 39(12):1317–1323, 1998

Niemi M, Backman JT, Neuvonen M, et al: Effects of fluconazole and fluvoxamine on the pharmacokinetics and pharmacodynamics of glimepiride. Clin Pharmacol Ther 69(4):194–200, 2001

Rasmussen BB, Nielsen TL, Brosen K: Fluvoxamine inhibits the CYP2C19-catalysed metabolism of proguanil in vitro. Eur J Clin Pharmacol 54(9–10):735–740, 1998

How the Mighty Have Fallen

A 56-year-old teacher with a seizure disorder had been virtually seizure-free for the past 10 years while taking phenytoin (Dilantin), 300 mg/day (blood level=14.5 μg/mL). One summer, he received the news that he would not be rehired by the school for the coming academic year. This event, along with the recent death of his spouse from pancreatic cancer, led him into his first severe depressive episode. His friends advised him to visit a psychiatrist, and he accepted their advice. During the intake, he ruminated about how he had attended the same school as a teenager, and in his day he had been the class president and the starting quarterback. Now he was being rejected by a place he had considered a kind of home. His obsessional ruminations about his "lost glories," as well as a full array of neurovegetative depressive symptoms, led the psychiatrist to start the patient on fluvoxamine (Luvox), with the plan to titrate to a dosage of 150 mg/day and then wait a few weeks for a possible response. During this fluvoxamine dose titration, the patient felt progressively more sedated and slightly unsteady, but he did not want to be a "complainer," and he decided to just "stick it out," assuming that these were transient side effects of his new medicine that would soon abate. However, 1 week after reaching 150 mg/day, the patient lost his balance and fell down a flight of stairs in his home. He was too delirious and debilitated to summon help himself, and he basically lay there until a friend happened to drop by 6 hours later. He was immediately taken to the local ER, where his phenytoin blood level was found to be 46.7 μg/mL (Mamiya et al. 2001).

Discussion

This is an example of an inhibitor added to a substrate.

Phenytoin is mostly a substrate of 2C9 and 2C19 (Cadle et al. 1994; Mamiya et al. 1998), and fluvoxamine is a strong inhibitor of 1A2, 2C9, and 2C19 and a moderate inhibitor of 3A4 (Christensen et al. 2002; Niemi et al. 2001; von Moltke et al. 1995). Thus, the addition of the fluvoxamine impaired the ability of 2C9 and 2C19 to significantly contribute to the metabolism of phenytoin, and this caused a more than threefold increase in the phenytoin blood level. The patient's state of phenytoin toxicity culminated in his delirium and subsequent fall.

References

Cadle RM, Zenon GJ 3rd, Rodriguez-Barradas MC, et al: Fluconazole-induced symptomatic phenytoin toxicity. Ann Pharmacother 28(2):191–195, 1994

Christensen M, Tybring G, Mihara K, et al: Low daily 10-mg and 20-mg doses of fluvoxamine inhibit the metabolism of both caffeine (cytochrome P4501A2) and omeprazole (cytochrome P4502C19). Clin Pharmacol Ther 71:141–152, 2002

Mamiya K, Ieiri I, Shimamoto J, et al: The effects of genetic polymorphisms of CYP2C9 and CYP2C19 on phenytoin metabolism in Japanese adult patients with epilepsy: studies in stereoselective hydroxylation and population pharmacokinetics. Epilepsia 39(12):1317–1323, 1998

Mamiya K, Kojima K, Yukawa E, et al: Phenytoin intoxication induced by fluvoxamine. Ther Drug Monit 23(1):75–77, 2001

Niemi M, Backman JT, Neuvonen M, et al: Effects of fluconazole and fluvoxamine on the pharmacokinetics and pharmacodynamics of glimepiride. Clin Pharmacol Ther 69(4):194–200, 2001

von Moltke LL, Greenblatt DJ, Court MH, et al: Inhibition of alprazolam and desipramine hydroxylation in vitro by paroxetine and fluvoxamine: comparison with other selective serotonin reuptake inhibitor antidepressants. J Clin Psychopharmacol 15(2):125–131, 1995

Double Fault

A 23-year-old professional tennis player from South Africa was visiting the United States in order to play in a tennis tournament. She visited a physician in the host city because she was troubled by increasingly unpleasant vaginal itching and burning. The physician asked her if she was taking any medications at present. She replied that she took a medication called Epanutin because she sometimes had "fits." Because of her accent, the physician thought she was taking this medication as a way of maintaining her high fitness level. She was eventually diagnosed with vaginal candidiasis, and the physician prescribed fluconazole (Diflucan), 200 mg on day 1 and 100 mg/day for 2 weeks thereafter. By day 5 of the fluconazole, however, she was pervasively groggy, uncoordinated, nauseated, and slurring her speech. After determining that she had not just returned from a drinking binge, her coach arranged for her to be taken to the nearest ER. He also brought her medication bottles with them so that the ER physician was able to learn that Epanutin was phenytoin (Dilantin in the United States). She was taking 300 mg/day, but her current phenytoin blood level was 32.8 μg/mL.

Discussion

This is an example of an inhibitor added to a substrate.

Phenytoin is mostly a substrate of 2C9 and 2C19 (Cadle et al. 1994; Mamiya et al. 1998), and fluconazole is a strong 2C9 inhibitor (Niemi et al. 2001). Thus, the addition of the fluconazole significantly impaired the ability of 2C9 to contribute to the metabolism of phenytoin. Since 2C19 was not able to compensate for the inhibition of 2C9, this led to an increase in the blood level of phenytoin and the patient's signs of phenytoin toxicity (Blum et al. 1991).

References

Blum RA, Wilton JH, Hilligoss DM, et al: Effect of fluconazole on the disposition of phenytoin. Clin Pharmacol Ther 49(4):420–425, 1991

Cadle RM, Zenon GJ 3rd, Rodriguez-Barradas MC, et al: Fluconazole-induced symptomatic phenytoin toxicity. Ann Pharmacother 28(2):191–195, 1994

Mamiya K, Ieiri I, Shimamoto J, et al: The effects of genetic polymorphism of CYP2C9 and CYP2C19 on phenytoin metabolism in Japanese adult patients with epilepsy: studies in stereoselective hydroxylation and population pharmacokinetics. Epilepsia 39(12):1317–1323, 1998

Niemi M, Backman JT, Neuvonen M, et al: Effects of fluconazole and fluvoxamine on the pharmacokinetics and pharmacodynamics of glimepiride. Clin Pharmacol Ther 69(4):194–200, 2001

Nystagmus

A 32-year-old woman with a long-standing seizure disorder was doing well with phenytoin (Dilantin), 300 mg/day (blood level=17.3 µg/mL). However, she had been experiencing symptoms consistent with premenstrual dysphoric disorder for the past year, and she eventually made an appointment to discuss this with her gynecologist. Her gynecologist opted to start her on fluoxetine (Sarafem, in this case), 20 mg/day, to address this issue. She tolerated the fluoxetine with no appreciable side effects. However, she began to experience some new sedation within 5 days of starting the fluoxetine. Two days later, she was extremely groggy, dizzy, and nauseated and was seeing double. She had a friend transport her to the nearest ER, where she was grossly ataxic and displayed vertical nystagmus on her neurological examination. A phenytoin level was 30.3 µg/mL (K.L. Cozza, S.C. Armstrong, personal communication, May 2002).

Discussion

This is an example of an inhibitor added to a substrate.

Phenytoin is mostly a substrate of 2C9 and 2C19 (Cadle et al. 1994; Mamiya et al. 1998). Fluoxetine (in concert with its active metabolite, norfluoxetine) is a strong inhibitor of 2D6 and a moderate inhibitor of 2C9, 2C19, and 3A4 (Greenblatt et al. 1999; Stevens and Wrighton 1993). Thus, the addition of fluoxetine significantly impaired the ability of 2C9 and 2C19 to significantly contribute to the metabolism of phenytoin, leading to an increase in the phenytoin blood level, even though the phenytoin dosage had not been changed (Nelson et al. 2001; Nightingale 1994). This increase in the phenytoin blood level was not as great in magnitude as it would have been with fluvoxamine, but it was enough to produce significant signs of phenytoin toxicity.

References

Cadle RM, Zenon GJ 3rd, Rodriguez-Barradas MC, et al: Fluconazole-induced symptomatic phenytoin toxicity. Ann Pharmacother 28(2):191–195, 1994

Greenblatt DJ, von Moltke LL, Harmatz JS, et al: Human cytochromes and some newer antidepressants: kinetics, metabolism, and drug interactions. J Clin Psychopharmacol 19 (5, suppl 1):23S–35S, 1999

Mamiya K, Ieiri I, Shimamoto J, et al: The effects of genetic polymorphisms of CYP2C9 and CYP2C19 on phenytoin metabolism in Japanese adult patients with epilepsy: studies in stereoselective hydroxylation and population pharmacokinetics. Epilepsia 39(12):1317–1323, 1998

Nelson MH, Birnbaum AK, Remmel RP: Inhibition of phenytoin hydroxylation in human liver microsomes by several selective serotonin re-uptake inhibitors. Epilepsy Res 44(1):71–82, 2001

Nightingale SL: From the Food and Drug Administration. JAMA 271(14):1067, 1994

Stevens JC, Wrighton SA: Interaction of the enantiomers of fluoxetine and norfluoxetine with human liver cytochromes P450. J Pharmacol Exp Ther 266(2): 964–971, 1993

Transitions

A 17-year-old with major depressive disorder was responding well to fluoxetine (Prozac), 20 mg/day. He then had his first seizure, resulting in a hospital admission, a full neurological workup, and the eventual diagnosis of a seizure disorder (which his mother had). His neurologist planned to prescribe phenytoin (Dilantin), but the child and adolescent psychiatry fellow who was treating the patient for his depression took the initiative to contact the neurologist and share his understanding that the fluoxetine would impair the metabolism of phenytoin. This meant that dosing of the phenytoin should be slower than normal and that blood levels should be checked at a lower dosage than would be typical. The neurologist appreciated the input, and in collaboration with the fellow he carefully titrated the dosage of phenytoin to 160 mg/day (blood level = 14.1 μg/mL).

The patient did well after the setback, until it was time for the fellow to relocate and a new child and adolescent psychiatry fellow assumed treatment of the patient. Two months into that treatment, the patient mentioned that he had long noticed that the fluoxetine impaired his ability to ejaculate, and he asked if the fellow would consider switching him to another antidepressant. Since the fellow was not as conversant with P450 issues as his predecessor, he readily agreed and helped orchestrate a crossover titration from fluoxetine to mirtazapine (Remeron). This crossover proceeded uneventfully, but 1 month later the patient had another seizure. When he was taken to the ER, his phenytoin blood level was found to be only 6.6 μg/mL.

Discussion

This is an example of reversal of inhibition.

Phenytoin is mostly a substrate of 2C9 and 2C19 (Cadle et al. 1994; Mamiya et al. 1998). Fluoxetine (in concert with its active metabolite, norfluoxetine) is a strong inhibitor of 2D6 and a moderate inhibitor of 2C9, 2C19, and 3A4 (Greenblatt et al. 1999; Stevens and Wrighton 1993). As detailed above, the first psychiatry fellow recognized that the addition of phenytoin to fluoxetine would follow the "substrate added to inhibitor" paradigm. Since 2C9 and 2C19 were significantly inhibited, their ability to efficiently metabolize phenytoin was impaired. Therefore, any amount of phenytoin that was to be added would be expected to generate a higher blood level than would have been achieved had the fluoxetine not been present (Nelson et al. 2001; Nightingale 1994). Once the final phenytoin dosage of 160 mg/day was generating a stable therapeutic blood level, the maintenance of this therapeutic blood level, at this dosage of phenytoin, was dependent on the continued presence of the 2C9/2C19 inhibitor, fluoxetine. Discontinuation of the fluoxetine by the new child and adolescent psychiatry fellow in the course of the crossover to mirtazapine allowed 2C9 and 2C19 to resume their baseline, uninhibited levels of activity, with the result that the phenytoin level then fell below the therapeutic range and the patient experienced another seizure.

References

Cadle RM, Zenon GJ 3rd, Rodriguez-Barradas MC, et al: Fluconazole-induced symptomatic phenytoin toxicity. Ann Pharmacother 28(2):191–195, 1994

Greenblatt DJ, von Moltke LL, Harmatz JS, et al: Human cytochromes and some newer antidepressants: kinetics, metabolism, and drug interactions. .J Clin Psychopharmacol 19 (5, suppl 1):23S–35S, 1999

Mamiya K, Ieiri I, Shimamoto J, et al: The effects of genetic polymorphisms of CYP2C9 and CYP2C19 on phenytoin metabolism in Japanese adult patients with epilepsy: studies in stereoselective hydroxylation and population pharmacokinetics. Epilepsia 39(12):1317–1323, 1998

Nelson MH, Birnbaum AK, Remmel RP: Inhibition of phenytoin hydroxylation in human liver microsomes by several selective serotonin re-uptake inhibitors. Epilepsy Res 44(1):71–82, 2001

Nightingale SL: From the Food and Drug Administration. JAMA 271(4):1067, 1994

Stevens JC, Wrighton SA: Interaction of the enantiomers of fluoxetine and norfluoxetine with human liver cytochromes P450. J Pharmacol Exp Ther 266(2): 964–971, 1993

From Heartbreak to Heartburn

A 58-year-old widower was chronically had been taking phenytoin (Dilantin), 300 mg bid, for a seizure disorder. His most recent phenytoin blood level was 16.1 μg/mL. He had recently broken up with a woman he had been dating for 18 months, and he was quite demoralized in the wake of the breakup. He coped with this loss by increasing his food intake, especially in the late evenings. In addition to gaining several pounds, he also experienced a worsening of heartburn pain, which he had felt off and on for the past 5 years. He eventually made an appointment with his internist, who prescribed omeprazole (Prilosec), 40 mg/day. Within several days, the patient began to experience increased sedation and a sensation of being less mentally acute. After 1 week on the omeprazole, he was frankly unsteady on his feet and felt drunk. He contacted his neurologist and informed him of both his current symptoms, as well as of the fact that he was now taking omeprazole. The neurologist told him to call 911 for an ambulance to transport him to the nearest ER, where his phenytoin blood level was found to be 27.9 μg/mL.

Discussion

This is a mixed example of a P450 inhibitor (omeprazole) added to a substrate (phenytoin) and a P-glycoprotein inhibitor (omeprazole) added to a P-glycoprotein substrate (phenytoin).

First, phenytoin is mostly a substrate of 2C9 and 2C19 (Cadle et al. 1994; Mamiya et al. 1998), and omeprazole is a strong inhibitor of 2C19, as well as an inducer of 1A2 (Furuta et al. 2001; Nousbaum et al. 1994). Thus, the addition of omeprazole significantly impaired the ability of 2C19 to contribute to the efficient metabolism of phenytoin. The enzyme 2C9 was not able to fully compensate for this 2C19 inhibition, so that the inhibition of 2C19 by omeprazole caused the phenytoin blood level to rise into the toxic range, resulting in symptoms of clinical phenytoin toxicity (Gugler and Jensen 1985; Prichard et al. 1987).

Part of the increase in the phenytoin was likely attributable to the fact that phenytoin is a P-glycoprotein substrate (Weiss et al. 2001) and omeprazole is a P-glycoprotein inhibitor (Pauli-Magnus et al. 2001). Thus, the addition of the omeprazole inhibited the functioning of this transporter, such that phenytoin was less likely to be extruded from enterocytes back into the gut lumen, where it would have been excreted rather than absorbed. Instead, the inhibition of the P-glycoprotein transporter allowed more phenytoin to remain in enterocytes, so that an increased amount was absorbed, ultimately resulting in an increase in the blood level of phenytoin.

References

Cadle RM, Zenon GJ 3rd, Rodriguez-Barradas MC, et al: Fluconazole-induced symptomatic phenytoin toxicity. Ann Pharmacother 28(2):191–195, 1994

Furuta S, Kamada E, Suzuki T, et al: Inhibition of drug metabolism in human liver microsomes by nizatidine, cimetidine and omeprazole. Xenobiotica 31(1):1–10, 2001

Gugler R, Jensen JC: Omeprazole inhibits oxidative drug metabolism: studies with diazepam and phenytoin in vivo and 7-ethoxycoumarin in vitro. Gastroenterology 89(6):1235–1241, 1985

Mamiya K, Ieiri I, Shimamoto J, et al: The effects of genetic polymorphism of CYP2C9 and CYP2C19 on phenytoin metabolism in Japanese adult patients with epilepsy: studies in stereoselective hydroxylation and population pharmacokinetics. Epilepsia 39(12):1317–1323, 1998

Nousbaum JB, Berthou F, Carlhant D, et al: Four-week treatment with omeprazole increases the metabolism of caffeine. Am J Gastroenterol 89(3):371–375, 1994

Pauli-Magnus C, Rekersbrink S, Klotz U, et al: Interaction of omeprazole, lansoprazole and pantoprazole with P-glycoprotein. Naunyn Schmiedebergs Arch Pharmacol 364(6):551–557, 2001

Prichard PJ, Walt RP, Kitchingman GK, et al: Oral phenytoin pharmacokinetics during omeprazole therapy. Br J Clin Pharmacol 24(4):453–455, 1987

Weiss ST, Silverman EK, Palmer LJ: Case-control association studies in pharmacogenetics. Pharmacogenetics 1(3):157–158, 2001

Clots

A 45-year-old woman with a seizure disorder had been successfully treated with phenytoin (Dilantin), 300 mg/day (blood levels generally around 16 μg/mL), for the past 5 years. One morning, she awoke with tingling on the right side of her face, and her speech was slightly slurred. She was taken to a local ER and eventually diagnosed with a transient ischemic attack. The treating physician decided to start the patient on ticlopidine (Ticlid), 250 mg bid. Within 1 week, the patient was severely somnolent, dizzy, and nauseated with vomiting. Fearing a full stroke, she called 911 and was again transported to the ER. Her computed tomography scan was unremarkable, but her phenytoin blood level was 39.0 μg/mL (Donahue et al. 1999).

Discussion

This is an example of an inhibitor added to a substrate.

Phenytoin is mostly a substrate of 2C9 and 2C19 (Cadle et al. 1994; Mamiya et al. 1998), and ticlopidine is a strong inhibitor of 2C19 (Ko et al. 2000). Thus, the addition of ticlopidine significantly impaired the ability of 2C19 to contribute to the metabolism of phenytoin. Since 2C9 was not able to compensate for ticlopidine's strong 2C19 inhibition, there was an increase in the blood level of phenytoin, even though the phenytoin dosage had not been increased. This resulted in a state of clinical phenytoin toxicity.

References

Cadle RM, Zenon GJ 3rd, Rodriguez-Barradas MC, et al: Fluconazole-induced symptomatic phenytoin toxicity. Ann Pharmacother 28(2):191–195, 1994

Donahue S, Flockhart DA, Abernathy DR: Ticlopidine inhibits phenytoin clearance. Clin Pharmacol Ther 66(6):563–568, 1999

Ko JW, Desta Z, Soukhova NV, et al: In vitro inhibition of the cytochrome P450 (CYP450) system by the antiplatelet drug ticlopidine: potent effect on CYP2C19 and CYP2D6. Br J Clin Pharmacol 49(4):343–351, 2000

Mamiya K, Ieiri I, Shimamoto J, et al: The effects of genetic polymorphism of CYP2C9 and CYP2C19 on phenytoin metabolism in Japanese adult patients with epilepsy: studies in stereoselective hydroxylation and population pharmacokinetics. Epilepsia 39(12):1317–1323, 1998

Dry Delirium

A 48-year-old man with alcohol dependence had just been admitted to the medical floor of a hospital for the treatment of active delirium tremens (DTs). The patient had already had a seizure in the ER, and after the patient was acutely stabilized, the hospitalist placed him on phenytoin (Dilantin), 400 mg/day (blood level=16.2 μg/mL), and a lorazepam (Ativan) taper to address his DTs. A psychiatric consultant visited the patient, who seemed willing to make a serious attempt to curtail his drinking. He reported to the psychiatrist that he had found receiving disulfiram (Antabuse) to be a helpful way to motivate him to stay "dry." The psychiatrist agreed to start him on disulfiram, 500 mg/day. Five days later, the patient seemed intoxicated and irritable, which led the hospital staff to believe that he had somehow procured alcohol while on one of his brief trips to the ER entrance to smoke a cigarette. However, he had been compliant with the disulfiram, and there was no evidence of a disulfiram-alcohol reaction. A Breathalyzer borrowed from the ER confirmed that he was still "dry." A stat phenytoin blood level was then ordered and found to be 31.9 μg/mL.

Discussion

This is an example of an inhibitor added to a substrate.

Phenytoin is mostly a substrate of 2C9 and 2C19 (Cadle et al. 1994; Mamiya et al. 1998). Disulfiram is certainly a strong 2E1 inhibitor, and it has been shown to increase phenytoin blood levels, but beyond that its enzymatic activities have not been well characterized (Emery et al. 1999; Kharasch et al. 1999; Kiorboe 1966). Disulfiram has been shown to also increase blood levels of diazepam (Valium) (metabolized primarily by 2C19 and secondarily by 3A4) (MacLeod et al. 1978; Ono et al. 1996). It has also been shown to *not* increase levels of alprazolam (Xanax) (metabolized by 3A4) (Diquet et al. 1990; Dresser et al. 2000). Disulfiram also has no known inhibitory profile with regard to phase II glucuronidation or P-glycoprotein systems. In view of this evidence, my belief is that, at the very least, disulfiram acts as an inhibitor of 2C19 as well as of 2E1. Thus, in a manner similar to ticlopidine (Ticlid) (see "Clots," p. 145), disulfiram's presumed strong 2C19 inhibition significantly impaired the ability of this enzyme to contribute to the metabolism of phenytoin, resulting in an increase in the phenytoin blood level and the accompanying delirium. Since disulfiram's inhibitory profile is otherwise not well characterized, it may be that 2C9 inhibition could also have played a role, but this is merely a possibility that cannot be excluded as yet, and perhaps not even a likely one.

References

Cadle RM, Zenon GJ 3rd, Rodriguez-Barradas MC, et al: Fluconazole-induced symptomatic phenytoin toxicity. Ann Pharmacother 28(2):191–195, 1994

Diquet B, Gujadhur L, Lamiable D, et al: Lack of interaction between disulfiram and alprazolam in alcoholic patients. Eur J Clin Pharmacol 38(2):157–160, 1990

Dresser GK, Spence JD, Bailey DG: Pharmacokinetic-pharmacodynamic consequences and clinical relevance of cytochrome P450 3A4 inhibition. Clin Pharmacokinet 38(1):41–57, 2000

Emery MG, Jubert C, Thummel KE, et al: Duration of cytochrome P-450 2E1 (CYP2E1) inhibition and estimation of functional CYP2E1 enzyme half-life after single-dose disulfiram administration in humans. J Pharmacol Exp Ther 291(1):213–219, 1999

Kharasch ED, Hankins DC, Jubert C, et al: Lack of single-dose disulfiram effects on cytochrome P-450 2C9, 2C19, 2D6, and 3A4 activities: evidence for specificity toward P-450 2E1. Drug Metab Dispos 27(6):717–723, 1999

Kiorboe E: Phenytoin intoxication during treatment with Antabuse (disulfiram). Epilepsia 7(3):246–249, 1966

MacLeod SM, Sellers EM, Giles HG, et al: Interaction of disulfiram with benzodiazepines. Clin Pharmacol Ther 24(5):583–589, 1978

Mamiya K, Ieiri I, Shimamoto J, et al: The effects of genetic polymorphisms of CYP2C9 and CYP2C19 on phenytoin metabolism in Japanese adult patients with epilepsy: studies in stereoselective hydroxylation and population pharmacokinetics. Epilepsia 39(12):1317–1323, 1998

Ono S, Hatanaka T, Miyazawa S, et al: Human liver microsomal diazepam metabolism using cDNA-expressed cytochrome P450s: role of CYP2B6, 2C19 and the 3A subfamily. Xenobiotica 26(11):1155–1166, 1996

Alcohol-Free Intoxication

A 35-year-old woman with a history of alcohol dependence and panic disorder had been taking disulfiram (Antabuse), 250 mg/day, for the past 2 years, and this helped her to remain abstinent from alcohol for this period. However, various stressors in her life had led to a resurgence of debilitating panic attacks. Her panic attacks had responded to diazepam (Valium), 5 mg tid, in the past, and her psychiatrist agreed to prescribe this until the sertraline (Zoloft), 100 mg/day, that he was also starting (initial dosage=25 mg/day) began to provide some relief. However, after 4 days the patient reported that she had had no more panic attacks but also that she felt much more sedated and almost "drunk," as compared with her prior response to this dosage of diazepam. The psychiatrist decreased the diazepam dosage to 2 mg tid, which caused a remission of the oversedation but some return of panic attacks, although not nearly as badly as before the diazepam was initiated. The psychiatrist instructed the patient to use a pill cutter to halve the diazepam 2-mg tablets, and then to take 3 mg tid, which provided optimal relief from panic without oversedation.

Discussion

This is an example of a substrate added to an inhibitor.

Diazepam is a substrate primarily of 2C19 and secondarily of 3A4 (Ono et al. 1996). Disulfiram is certainly a strong 2E1 inhibitor, and it has been shown to increase phenytoin blood levels, but beyond that its enzymatic activities have not been well characterized (Emery et al. 1999; Kharasch et al. 1999; Kiorboe 1966). Disulfiram has also been shown to increase blood levels of diazepam (Valium) (metabolized primarily by 2C19 and secondarily by 3A4) (MacLeod et al. 1978; Ono et al. 1996). Also, it has been shown to *not* increase levels of alprazolam (Xanax) (metabolized by 3A4) (Diquet et al. 1990; Dresser et al. 2000). Disulfiram also has no known inhibitory profile with regard to phase II glucuronidation or P-glycoprotein systems. In view of this evidence, my belief is that, at the very least, disulfiram acts as an inhibitor of 2C19 as well as of 2E1. This is the likely mechanism by which disulfiram elevates the blood level of diazepam.

This explains why the patient became oversedated while taking a dosage of diazepam that she had previously tolerated. The psychiatrist compensated for this effect by carefully titrating, on the basis of clinical effect, to a 40% reduction of the diazepam dosage.

References

Diquet B, Gujadhur L, Lamiable D, et al: Lack of interaction between disulfiram and alprazolam in alcoholic patients. Eur J Clin Pharmacol 38(2):157–160, 1990

Dresser GK, Spence JD, Bailey DG: Pharmacokinetic-pharmacodynamic consequences and clinical relevance of cytochrome P450 3A4 inhibition. Clin Pharmacokinet 38(1):41–57, 2000

Emery MG, Jubert C, Thummel KE, et al: Duration of cytochrome P-450 2E1 (CYP2E1) inhibition and estimation of functional CYP2E1 enzyme half-life after single-dose disulfiram administration in humans. J Pharmacol Exp Ther 291(1):213–219, 1999

Kharasch ED, Hankins DC, Jubert C, et al: Lack of single-dose disulfiram effects on cytochrome P-450 2C9, 2C19, 2D6, and 3A4 activities: evidence for specificity toward P-450 2E1. Drug Metab Dispos 27(6):717–723, 1999

Kiorboe E: Phenytoin intoxication during treatment with Antabuse (disulfiram). Epilepsia 7(3):246–249, 1966

MacLeod SM, Sellers EM, Giles HG, et al: Interaction of disulfiram with benzodiazepines. Clin Pharmacol Ther 24(5):583–589, 1978

Ono S, Hatanaka T, Miyazawa S, et al: Human liver microsomal diazepam metabolism using cDNA-expressed cytochrome P450s: role of CYP2B6, 2C19 and the 3A subfamily. Xenobiotica 26(11):1155–1166, 1996

Just Desserts

A 74-year-old man with bipolar I disorder and type II diabetes mellitus was hospitalized due to a manic episode. During the initial phase of his inpatient stay, he refused all psychiatric medications, but he readily took his prescribed glipizide (Glucotrol), 2.5 mg/day. However, he did refuse to abide by a diabetic diet, stating, “Since I'm going to eat whatever I want after I leave the hospital, why restrict my intake now?” The psychiatrist agreed to liberalize his diet, so long as he ate more than just the desserts on his and his peers' trays, which had been his practice to date. His glipizide was then titrated to 5 mg/day, and at that dosage he maintained consistent blood glucose readings between 120 and 170 mg/dL. After these and other maneuvers fostered the emergence of a therapeutic alliance, the patient agreed to a trial of divalproex sodium (Depakote), titrated to a dosage of 500 mg/day. After 5 days of taking the divalproex, however, he reported feeling severely light-headed, sweaty, and nervous. He was slightly tachycardic (110 beats per minute) but not at all orthostatic. The nurses checked the glucose level by fingerstick, which was found to be 51 mg/dL (S.C. Armstrong, personal communication, May 2002).

Discussion

This is an example of an inhibitor added to a substrate.

Glipizide is a 2C9 substrate (Kidd et al. 2001), and divalproex is a 2C9 inhibitor (Wen et al. 2001), among having many other metabolic roles. The addition of the divalproex led to a significant impairment in the ability of 2C9 to efficiently metabolize the glipizide. Consequently, the blood level of glipizide rose, even though the glipizide dosage had not been further increased at that time. Since glipizide is an oral hypoglycemic agent, an increase in the glipizide blood level led to a decrease in the patient's blood glucose, with accompanying clinical signs of hypoglycemia. This would probably be best addressed by switching to another hypoglycemic agent whose metabolism would not be affected by divalproex, but another short-term option would be to decrease the dosage of glipizide.

References

Kidd RS, Curry TB, Gallagher S, et al: Identification of a null allele of *CYP2C9* in an African-American exhibiting toxicity to phenytoin. Pharmacogenetics 11(9):803–808, 2001

Wen X, Wang JS, Kivisto KT, et al: In vitro evaluation of valproic acid as an inhibitor of human cytochrome P450 isoforms: preferential inhibition of cytochrome P450 2C9 (CYP2C9). Br J Clin Pharmacol 52(5):547–553, 2001

The Anxious Accountant (I)

A 55-year-old CPA with generalized anxiety disorder had responded well for 20 years to treatment with diazepam (Valium), 5 mg tid. Predictably, she always became more nervous around March because of the crush of yearly tax submissions. Her psychiatrist would supply her with one or two prn doses per week of diazepam to offset this period of heightened stress, and her use of diazepam had remained stable over the years. One March, she experienced increasing indigestion and heartburn pain, which prompted a visit to her internist. He prescribed omeprazole (Prilosec), 20 mg/day, for this problem. Within 5 days, she experienced an unusual absence of inner turmoil during what was usually a stressful period.

She was more efficient and unflustered during a tax season than she could ever recall, and her co-workers supported this perception. She did not require any extra doses of diazepam for the entire tax season. She shared this information with her psychiatrist, who was puzzled but pleasantly so. For his part, he had been surprised that he had not received any extra phone calls from the patient for the past month, but he ascribed that to character development and the patient's finding more mature coping mechanisms with age. She continued to enjoy a sense of calm that persisted beyond the tax season.

Discussion

This is an example of an inhibitor added to a substrate.

Diazepam is a substrate primarily of 2C19 and secondarily of 3A4 (Ono et al. 1996), and omeprazole is a strong inhibitor of 2C19, as well as an inducer of 1A2 (Furuta et al. 2001; Nousbaum et al. 1994). Thus, the addition of the omeprazole significantly impaired the ability of 2C19 to contribute to the metabolism of diazepam. Since 3A4 was not able to entirely compensate for this 2C19 inhibition, there was an increase in the blood level of diazepam.

Studies have demonstrated that the addition of omeprazole to diazepam will typically increase blood levels of diazepam by roughly 40% (Caraco et al. 1995). This increase in the diazepam blood level might have ordinarily produced unwanted sedation or cognitive slowing, but it appears that this patient had been chronically and mildly undertreated, such that the increase seems to have produced an as yet unrecognized clinical benefit for this patient.

References

Caraco Y, Tateishi T, Wood AJ: Interethnic difference in omeprazole's inhibition of diazepam metabolism. Clin Pharmacol Ther 58(1):62–72, 1995

Furuta S, Kamada E, Suzuki T, et al: Inhibition of drug metabolism in human liver microsomes by nizatidine, cimetidine and omeprazole. Xenobiotica 31(1):1–10, 2001

Nousbaum JB, Berthou F, Carlhant D, et al: Four-week treatment with omeprazole increases the metabolism of caffeine. Am J Gastroenterol 89(3):371–375, 1994

Ono S, Hatanaka T, Miyazawa S, et al: Human liver microsomal diazepam metabolism using cDNA-expressed cytochrome P450s: role of CYP2B6, 2C19 and the 3A subfamily. Xenobiotica 26(11):1155–1166, 1996

The Anxious Accountant (II)

The patient in the previous case continued to do well for several months, until the omeprazole was taken off her insurance company's formulary. Her internist replaced this with pantoprazole (Protonix). Within a few days, she experienced what felt like her "old" anxieties and worries, but no acute stressors were present. The lack of any such stressors prompted the psychiatrist to ask more questions, specifically about medication changes. When she revealed the recent change from omeprazole to pantoprazole, the psychiatrist did some quick research and then decided to increase her standing dose of diazepam to 7 mg tid. Within another 5 days, she felt like her "new" old self again.

Discussion

This is an example of reversal of inhibition (with a happy ending).

As mentioned in the previous case ("The Anxious Accountant [I]"), diazepam is a substrate primarily of 2C19 and secondarily of 3A4 (Ono et al. 1996), and omeprazole is a strong inhibitor of 2C19, as well as an inducer of 1A2 (Furuta et al. 2001; Nousbaum et al. 1994). Thus, the initial addition of the omeprazole significantly impaired the ability of 2C19 to contribute to the metabolism of diazepam. Since 3A4 was not able to entirely compensate for this 2C19 inhibition, there was an increase in the blood level of diazepam, which actually proved helpful for this patient (Caraco et al. 1995). However, with the discontinuation of the omeprazole and replacement with a proton pump inhibitor (pantoprazole) that lacks a 2C19 inhibitory profile, 2C19 was able to return to its greater baseline level of activity. This led to more efficient metabolism of the diazepam and subsequent decline of the diazepam blood level back to its previous level, which was likely slightly suboptimal. The psychiatrist eventually recognized the subtle benefit that the presence of the omeprazole had conferred, and he successfully reproduced this by increasing the dose of diazepam by 40% (from 5 mg tid to 7 mg tid).

References

Caraco Y, Tateishi T, Wood AJ: Interethnic difference in omeprazole's inhibition of diazepam metabolism. Clin Pharmacol Ther 58(1):62–72, 1995

Furuta S, Kamada E, Suzuki T, et al: Inhibition of drug metabolism in human liver microsomes by nizatidine, cimetidine and omeprazole. Xenobiotica 31(1):1–10, 2001

Nousbaum JB, Berthou F, Carlhant D, et al: Four-week treatment with omeprazole increases the metabolism of caffeine. Am J Gastroenterol 89(3):371–375, 1994

Ono S, Hatanaka T, Miyazawa S, et al: Human liver microsomal diazepam metabolism using cDNA-expressed cytochrome P450s: role of CYP2B6, 2C19 and the 3A subfamily. Xenobiotica 26(11):1155–1166, 1996

Sensitive

A 40-year-old woman with a seizure disorder and bipolar I disorder was being maintained on the following regimen: olanzapine (Zyprexa), 10 mg qhs; clonazepam (Klonopin), 1 mg bid; and phenytoin (Dilantin), 400 mg/day (most recent blood level=12.2 μg/mL). However, she then required hospitalization for a manic episode. She was already unhappy about taking the olanzapine, having gained 8 pounds since she started that medication 1 year ago. Therefore, she did not want the olanzapine dosage further increased, and she rejected lithium and divalproex sodium (Depakote) out of hand as potential mood stabilizers. Her psychiatrist decided to add topiramate (Topamax), initially at 50 mg/day, with the plan to further titrate the dose as clinically indicated. After 3 days, the patient reported ataxia and drowsiness. A new phenytoin level was drawn, which was 17.7 μg/mL. The psychiatrist decreased the phenytoin dosage to 300 mg/day, and the phenytoin blood level then returned to its pre-topiramate value. Topiramate was retained in the regimen, and it was eventually titrated to a dosage of 250 mg/day, to which the patient's manic symptoms appeared to respond (K. Walters, L. Lin, personal communication, August 2002).

Discussion

This is an example of an inhibitor (topiramate) added to a substrate (phenytoin) and (probably) a substrate (topiramate) added to an inducer (phenytoin).

First, phenytoin is mostly a substrate of 2C9 and 2C19 (Cadle et al. 1994; Mamiya et al. 1998), and topiramate is an inhibitor of 2C19 (Anderson 1998) and an inducer of 3A4 (Benedetti 2000). Thus, the addition of topiramate significantly impaired the ability of 2C19 to contribute to the metabolism of phenytoin. Since the activity of 2C9 was not able to fully compensate, the inhibition of 2C19 by topiramate resulted in an increase in the blood level of phenytoin (Sachdeo et al. 2002)—in this case, by nearly 50%. The psychiatrist was able to compensate for this increase in the phenytoin blood level by decreasing the dosage from 400 mg/day to 300 mg/day.

Second, topiramate is a substrate of phase II glucuronidation (not well understood beyond that at present), although a significant portion of this compound is renally excreted (roughly 70%) and not hepatically metabolized at all (Topamax [package insert] 2000). Phenytoin is an inducer of 3A4, 2C9, 2C19, and glucuronidation enzyme 1A4 (Chetty et al. 1998; Rambeck et al. 1996; Spina and Perucca 2002). Studies have demonstrated that the addition of phenytoin will generally decrease the blood level of topiramate by 50% (Sachdeo et al. 2002). Therefore, the presence of the phenytoin likely necessitated a higher dose (probably roughly double) of topiramate to produce a clinical response than would have been needed in the absence of phenytoin. Since there is no current evidence that the P-glycoprotein transporter plays a role here, it stands to reason that topiramate is likely a substrate within the 1A family of phase II glucuronidation substrates.

It may seem odd that the patient reported symptoms consistent with phenytoin toxicity, even though her highest phenytoin blood level still fell well within the therapeutic range. This may be attributable to a relatively rapid change in the blood level, but it is more likely that this individual was simply idiosyncratically sensitive to phenytoin side effects at the upper portion of the therapeutic range.

References

Anderson GD: A mechanistic approach to antiepileptic drug interactions. Ann Pharmacother 32(5):554–563, 1998

Benedetti MS: Enzyme induction and inhibition by new antiepileptic drugs: a review of human studies. Fundam Clin Pharmacol 14(4):301–319, 2000

Cadle RM, Zenon GJ 3rd, Rodriguez-Barradas MC, et al: Fluconazole-induced symptomatic phenytoin toxicity. Ann Pharmacother 28(2):191–195, 1994

Chetty M, Miller R, Seymour MA: Phenytoin autoinduction. Ther Drug Monit 20(1):60–62, 1998

Mamiya K, Ieiri I, Shimamoto J, et al: The effects of genetic polymorphisms of CYP2C9 and CYP2C19 on phenytoin metabolism in Japanese adult patients with epilepsy: studies in stereoselective hydroxylation and population pharmacokinetics. Epilepsia 39(12):1317–1323, 1998

Rambeck B, Specht U, Wolf P: Pharmacokinetic interactions of the new antiepileptic drugs. Clin Pharmacokinet 31(4):309–324, 1996

Sachdeo RC, Sachdeo SK, Levy RH, et al: Topiramate and phenytoin pharmacokinetics during repetitive monotherapy and combination therapy to epileptic patients. Epilepsia 43(7):691–696, 2002

Spina E, Perucca E: Clinical significance of pharmacokinetic interactions between antiepileptic and psychotropic drugs. Epilepsia 43 (suppl 2):37–44, 2002

Topamax (package insert). Raritan, NJ, Ortho-McNeil Pharmaceutical Inc, May 2000

The Dizzy Dentist

A 63-year-old dentist had been taking phenytoin (Dilantin), 400 mg/day (blood level generally=13–15 μg/mL) for 25 years to treat his seizure disorder. He began to develop chronic indigestion pain, and he made an appointment with his internist to address this complaint. His internist prescribed ranitidine (Zantac), 150 mg bid. Over the course of the next week, the patient experienced increasing dizziness, difficulty concentrating, and tremor. These symptoms became so severe that he cancelled his workday for fear that he would hurt a patient by slipping during a procedure. He contacted his internist, who instructed the patient to report to the local ER for a stat phenytoin blood level, which was 21.8 μg/mL. His internist instructed him to immediately stop his ranitidine, and his symptoms remitted within 2 days of doing so.

Discussion

This is an example of an inhibitor added to a substrate.

Phenytoin is mostly a substrate of 2C9 and 2C19 (Cadle et al. 1994; Mamiya et al. 1998), and ranitidine is (probably) a mild to moderate inhibitor of both 2C9 and 2C19 (Bramhall and Levine 1988; Tse et al. 1993). Thus, the addition of ranitidine led to an impairment in the ability of 2C9 and 2C19 to contribute to the metabolism of phenytoin. This inhibition caused the blood level of phenytoin to increase, even though there had been no increase in the dosage of phenytoin (Tse et al. 1993).

Ranitidine is considered to pose minimal risks for drug interactions. It is certainly much less generally a culprit than the pan-inhibitor cimetidine (Tagamet). However, individual cases have demonstrated that the addition of ranitidine to phenytoin can increase phenytoin blood levels by roughly 40% (Bramhall and Levine 1988).

References

Bramhall D, Levine M: Possible interaction of ranitidine with phenytoin. Drug Intell Clin Pharmacol 22(12):979–980, 1988

Cadle RM, Zenon GJ 3rd, Rodriguez-Barradas MC, et al: Fluconazole-induced symptomatic phenytoin toxicity. Annals of Pharmacotherapy 28(2):191–195, 1994

Mamiya K, Ieiri I, Shimamoto J, et al: The effects of genetic polymorphisms of CYP2C9 and CYP2C19 on phenytoin metabolism in Japanese adult patients with epilepsy: studies in stereoselective hydroxylation and population pharmacokinetics. Epilepsia 39(12):1317–1323, 1998

Tse CS, Akinwande KI, Biallowons K: Phenytoin concentration elevation subsequent to ranitidine administration. Ann Pharmacother 27(12):1448–1451, 1993

Bruiser

A 75-year-old woman with a history of previous major depressive episodes (no treatment for years) was hospitalized on the medical floor following a pulmonary embolus due to formation of a deep venous thrombosis in her right thigh. She was initially treated with intravenous heparin and then transitioned to warfarin (Coumadin), 5 mg/day (International Normalized Ratio, or INR=2.4). After 3 weeks in the hospital, she experienced a recurrence of her major depressive disorder. To minimize the likelihood of plasma protein displacement with one of the more highly protein-bound selective serotonin reuptake inhibitors (paroxetine [Paxil], sertraline [Zoloft], or fluoxetine [Prozac]), her internist prescribed fluvoxamine (Luvox), titrating the medication upward to a dosage of 100 mg/day. After 1 week of taking this dosage of fluvoxamine, the patient accidentally, but lightly, bumped her arm on the side rail of her hospital bed. She was surprised that she immediately developed a large and ugly bruise on her arm as a result. She informed her internist about this, and he ordered a stat prothrombin time, which revealed an INR of 5.8. The fluvoxamine was immediately discontinued, but her INR did not fall below 3.0 for another 10 days.

Discussion

This is an example of an inhibitor added to a substrate.

Warfarin's metabolism is exceedingly complex, but it mostly occurs at 2C9 for the more active *S*-warfarin isomer (Heimark et al. 1987; Linder and Valdes 1999). Fluvoxamine is a strong inhibitor of 1A2, 2C9, and 2C19 and a moderate inhibitor of 3A4 (Christensen et al. 2002; Niemi et al. 2001; von Moltke et al. 1995). Thus, the addition of fluvoxamine significantly impaired the ability of 2C9 to metabolize the *S*-warfarin, leading to an increase in the blood level of this warfarin isomer, even though the dosage of warfarin had remained constant throughout (Yap and Low 1999). This caused a significant increase in the patient's INR, to almost 2½ times the baseline value, which placed the patient in a pathologically hypocoagulable state in which she bruised very easily and was in danger of vascular hemorrhage. The internist eventually had to administer vitamin K to address her persistently high INR.

References

Christensen M, Tybring G, Mihara K, et al: Low daily 10-mg and 20-mg doses of fluvoxamine inhibit the metabolism of both caffeine (cytochrome P4501A2) and omeprazole (cytochrome P4502C19). Clin Pharmacol Ther 71(4):141–152, 2002

Heimark LD, Gibaldi M, Trager WF, et al: The mechanism of the warfarin-rifampin drug interaction. Clin Pharmacol Ther 42:388–394, 1987

Linder MW, Valdes R Jr: Pharmacogenetics in the practice of laboratory medicine. Mol Diagn 4(4):365–379, 1999

Niemi M, Backman JT, Neuvonen M, et al: Effects of fluconazole and fluvoxamine on the pharmacokinetics and pharmacodynamics of glimepiride. Clin Pharmacol Ther 69(4):194–200, 2001

von Moltke LL, Greenblatt DJ, Court MH, et al: Inhibition of alprazolam and desipramine hydroxylation in vitro by paroxetine and fluvoxamine: comparison with other selective serotonin reuptake inhibitor antidepressants. J Clin Psychopharmacol 15(2):125–131, 1995

Yap KB, Low ST: Interaction of fluvoxamine with warfarin in an elderly woman. Singapore Med J 40(7):480–482, 1999

Sleeping Beauty

A 21-year-old woman was receiving diazepam (Valium), 5 mg bid, from her family physician for a complaint of "free-floating anxiety." With the beginnings of her first serious intimate relationship, she experienced the new onset of obsessions around order and compulsive checking. These symptoms prompted her to visit a psychiatrist, who made the diagnosis of obsessive-compulsive disorder (OCD). He started the patient on fluvoxamine (Luvox), 50 mg bid initially and titrated to an eventual dosage of 200 mg/day. After 1 week at this dosage, the patient reported that she was sleeping more than 11 hours each night and that she was still sleepy during the day. Although she was able to drive safely, she felt that her ability to concentrate in her college classes was diminished. She reported these difficulties to her psychiatrist, who then decreased the dose of diazepam to 2 mg bid. Her sedation remitted and her OCD symptoms continued to improve over the succeeding weeks.

Discussion

This is an example of an inhibitor added to a substrate.

Diazepam is primarily a substrate of 2C19, with a smaller contribution from 3A4 (Ono et al. 1996). Fluvoxamine is a strong inhibitor of 1A2, 2C9, and 2C19 and a moderate inhibitor of 3A4 (Christensen et al. 2002; Niemi et al. 2001; von Moltke et al. 1995). Thus, the addition of the fluvoxamine significantly impaired the ability of 2C19 and 3A4 to efficiently metabolize the diazepam, leading to an increase in the blood level of diazepam (Perucca et al. 1994). This increase in the diazepam blood level caused the patient to become sedated. The psychiatrist nicely compensated for fluvoxamine's inhibition of diazepam's metabolism by decreasing the dose of diazepam from 5 mg bid to 2 mg bid. Although the blood level of diazepam while the patient was taking 4 mg/day along with the fluvoxamine might not have been quite as high as the blood level was at 10 mg/day without fluvoxamine, the patient tolerated this lower dose of diazepam without difficulty. Her positive response was possibly due, in part, to some anxiolytic contribution (albeit not through direct GABA [γ-aminobutyric acid] agonism) from the fluvoxamine.

References

Christensen M, Tybring G, Mihara K, et al: Low daily 10-mg and 20-mg doses of fluvoxamine inhibit the metabolism of both caffeine (cytochrome P4501A2) and omeprazole (cytochrome P4502C19). Clin Pharmacol Ther 71:141–152, 2002

Niemi M, Backman JT, Neuvonen M, et al: Effects of fluconazole and fluvoxamine on the pharmacokinetics and pharmacodynamics of glimepiride. Clin Pharmacol Ther 69(4):194–200, 2001

Ono S, Hatanaka T, Miyazawa S, et al: Human liver microsomal diazepam metabolism using cDNA-expressed cytochrome P450s: role of CYP2B6, 2C19 and the 3A subfamily. Xenobiotica 26(11):1155–1166, 1996

Perucca E, Gatti G, Cipolla G, et al: Inhibition of diazepam metabolism by fluvoxamine: a pharmacokinetic study in normal volunteers. Clin Pharmacol Ther 56(5):471–476, 1994

von Moltke LL, Greenblatt DJ, Court MH, et al: Inhibition of alprazolam and desipramine hydroxylation in vitro by paroxetine and fluvoxamine: comparison with other selective serotonin reuptake inhibitor antidepressants. J Clin Psychopharmacol 15(2):125–131, 1995

Optimization

A 16-year-old with nonseminomatous testicular cancer had undergone several chemotherapeutic regimens, with only partial or no success. In the midst of his treatment, he developed seizures and he was placed on phenytoin (Dilantin), which did prevent further seizure activity. After this event, he was given another chemotherapeutic trial that included ifosfamide (Ifex). His oncologist was pessimistic, on the basis of the relative failure of the previous chemotherapeutic trials. However, the patient went into a full remission following this trial.

Discussion

This is an example of a substrate added to an inducer.

Ifosfamide is, to a significant degree, a pro-drug, whose metabolic products possess much greater antineoplastic potency than the parent compound (Kan et al. 2001). Phenytoin is an inducer of 2C9, 2C19, 3A4, and likely 2B6 as well (Chetty et al. 1998; Ducharme et al. 1997; Spina and Perucca 2002). The likely induction of 2B6 by phenytoin probably increased the amount of 2B6 that was available to convert ifosfamide from the less-active parent compound to the more potently antineoplastic compounds *(S)*-2-DCE-ifosfamide and *(S)*-3-DCE-ifosfamide. In this case, it was hypothesized that pretreatment with phenytoin led to the increased production of these more potent ifosfamide metabolites, which resulted in the patient's remission following several previous failed trials (Ducharme et al. 1997). Cyclophosphamide follows a metabolic path similar to that of ifosfamide (Williams et al. 1999).

References

Chetty M, Miller R, Seymour MA: Phenytoin autoinduction. Ther Drug Monit 20(1):60–62, 1998

Ducharme MP, Bernstein ML, Granvil CP, et al: Phenytoin-induced alteration in the N-dechloroethylation of ifosfamide stereoisomers. Cancer Chemother Pharmacol 40(6):531–533, 1997

Kan O, Griffiths L, Baban D, et al: Direct retroviral delivery of human cytochrome P450 2B6 for gene-directed enzyme prodrug therapy of cancer. Cancer Gene Ther 8(7):473–482, 2001

Spina E, Perucca E: Clinical significance of pharmacokinetic interactions between antiepileptic and psychotropic drugs. Epilepsia 43 (suppl 2):37–44, 2002

Williams ML, Wainer IW, Embree L, et al: Enantioselective induction of cyclophosphamide metabolism by phenytoin. Chirality 11(7):569–574, 1999

All Things in Excess

A 23-year-old man with alcohol dependence discontinued his alcohol intake as a prerequisite for moving in with his brother. After 2 days of abstinence, he did not enter florid DTs, but he did have a moderate degree of tachycardia and hypertension, such that he experienced a severe vascular headache. He attempted to "nuke" this headache by taking a bolus of 3,250 mg of acetaminophen (Tylenol). After another 2 days, he told his brother that he was feeling weak, had no appetite, and was experiencing nausea, vomiting, and pain under his ribs on the right side. His brother also noted that his skin had a yellowish tint. The brother drove the patient to the nearest ER, where he was diagnosed with acute acetaminophen-induced hepatitis (K.L. Cozza, S.C. Armstrong, personal communication, May 2002).

Discussion

This is an example of a substrate added to an induced enzyme, just after the inducer had been discontinued.

Acetaminophen is a substrate of 2E1 (Manyike et al. 2000), and chronic alcohol consumption is a potent inducer of 2E1 (Seitz and Csomos 1992). Thus, when the patient took the large bolus of acetaminophen, there was a greater than normal amount of 2E1 that was available to metabolize acetaminophen down this specific metabolic pathway. In effect, the metabolism of acetaminophen was "dragged" down the 2E1-mediated pathway to a greater than normal degree, by virtue of the greater availability, and thus activity, of 2E1. However, this excessive metabolism of acetaminophen by 2E1 led to the accumulation of a hepatotoxic metabolite, *N*-acetyl-*p*-benzoquinone imine, or NAPQI (Manyike et al. 2000).

Conventional treatment at this point would involve the administration of *N*-acetylcysteine (Mucomyst) to promote glutathione synthesis and thus neutralize this toxic metabolite. A less conventional, but likely helpful, approach would be to also administer a 2E1 inhibitor, such as disulfiram (Antabuse) (Emery et al. 1999; Kharasch et al. 1999) or watercress (Leclercq et al. 1998), and thus decrease the generation of the NAPQI metabolite.

References

Emery MG, Jubert C, Thummel KE, et al: Duration of cytochrome P-450 2E1 (CYP2E1) inhibition and estimation of functional CYP2E1 enzyme half-life after single-dose disulfiram administration in humans. J Pharmacol Exp Ther 291(1):213–219, 1999

Kharasch ED, Hankins DC, Jubert C, et al: Lack of single-dose disulfiram effects on cytochrome P-450 2C9, 2C19, 2D6, and 3A4 activities: evidence for specificity toward P-450 2E1. Drug Metab Dispos 27(6):717–723, 1999

Leclercq I, Desager JP, Horsmans Y: Inhibition of chlorzoxazone metabolism, a clinical probe for CYP2E1, by a single ingestion of watercress. Clin Pharmacol Ther 64(2):144–149, 1998

Manyike PT, Kharasch ED, Kalhorn TF, et al: Contribution of CYP2E1 and CYP3A to acetaminophen reactive metabolite formation. Clin Pharmacol Ther 67(3): 275–282, 2000

Seitz HK, Csomos G: Alcohol and the liver: ethanol metabolism and the pathomechanism of alcoholic liver damage (in Hungarian). Orv Hetil 133(50): 3183–3189, 1992

Antidepressant Withdrawal Seizure

A 22-year-old man with bipolar I disorder and a seizure disorder was being maintained on lithium, 900 mg/day (blood level=0.75 mEq/L); fluvoxamine (Luvox), 150 mg/day; and phenytoin (Dilantin), 300 mg/day (blood level=12.4 μg/mL). The phenytoin was the most recent addition to this regimen, as he had been taking a different anticonvulsant before he was switched to phenytoin. He had been seizure-free for the previous 2 years. The patient began to display emerging manic symptoms (decreasing need for sleep, increasing libido, increased rate of speech and distractibility), so his psychiatrist discontinued the fluvoxamine. Ten days later, the patient experienced a seizure and his phenytoin blood level was found to be only 4.7 μg/mL.

Discussion

This is an example of reversal of inhibition.

Phenytoin is mostly a substrate of 2C9 and 2C19 (Cadle et al. 1994; Mamiya et al. 1998), and fluvoxamine is a strong inhibitor of 1A2, 2C9, and 2C19 and a moderate inhibitor of 3A4 (Christensen et al. 2002; Niemi et al. 2001; von Moltke et al. 1995). Thus, the therapeutic blood level of phenytoin at a dosage of 300 mg/day relied on the inhibition of phenytoin's metabolism at 2C9 and 2C19 by fluvoxamine (Mamiya et al. 2001). When the fluvoxamine was discontinued, 2C9 and 2C19 resumed their higher baseline levels of activity, resulting in more efficient metabolism of the phenytoin and a subsequent decrease in the phenytoin blood level to the subtherapeutic range. This caused the patient to have his first seizure in 2 years. This problem was eventually addressed by the psychiatrist's increasing the phenytoin dosage to 600 mg/day, which produced a blood level of 10.8 μg/mL.

References

Cadle RM, Zenon GJ 3rd, Rodriguez-Barradas MC, et al: Fluconazole-induced symptomatic phenytoin toxicity. Ann Pharmacother 28(2):191–195, 1994

Christensen M, Tybring G, Mihara K, et al: Low daily 10-mg and 20-mg doses of fluvoxamine inhibit the metabolism of both caffeine (cytochrome P4501A2) and omeprazole (cytochrome P4502C19). Clin Pharmacol Ther 71:141–152, 2002

Mamiya K, Ieiri I, Shimamoto J, et al: The effects of genetic polymorphisms of CYP2C9 and CYP2C19 on phenytoin metabolism in Japanese adult patients with epilepsy: studies in stereoselective hydroxylation and population pharmacokinetics. Epilepsia 39(12):1317–1323, 1998

Mamiya K, Kojima K, Yukawa E, et al: Phenytoin intoxication induced by fluvoxamine. Ther Drug Monit 23(1):75–77, 2001

Niemi M, Backman JT, Neuvonen M, et al: Effects of fluconazole and fluvoxamine on the pharmacokinetics and pharmacodynamics of glimepiride. Clin Pharmacol Ther 69(4):194–200, 2001

von Moltke LL, Greenblatt DJ, Court MH, et al: Inhibition of alprazolam and desipramine hydroxylation in vitro by paroxetine and fluvoxamine: comparison with other selective serotonin reuptake inhibitor antidepressants. J Clin Psychopharmacol 15(2):125–131, 1995

More Is Less

A 40-year-old man with a history of seizures since age 10 had been receiving phenytoin (Dilantin), 100 mg tid (blood levels generally between 12 and 15 μg/mL) for the past 4 years, with only partial control of his illness. He continued to have breakthrough seizures once every 3–4 weeks. The patient and his neurologist decided to attempt to further optimize his treatment. The neurologist added carbamazepine (Tegretol), 100 mg bid for 4 days and then 200 mg bid thereafter. Starting 1 week later, the patient actually experienced a marked increase in the frequency of his seizures. After another week had passed, the patient was experiencing at least one seizure each day. He was eventually brought to the local ER by his family, where the blood level of phenytoin was found to be 6.6 µg/mL and the carbamazepine blood level was 2.4 µg/mL (D. Boerescu, personal communication, July 2002).

Discussion

This is a combined example of a substrate (carbamazepine) added to an inducer (phenytoin) and an inducer (carbamazepine) added to a substrate (phenytoin).

First, carbamazepine is primarily a 3A4 substrate, although 1A2 and 2C9 also make minor contributions to the metabolism of carbamazepine (Spina et al. 1996). Phenytoin is an inducer of 3A4 and 2C9 (Chetty et al. 1998; Spina and Perucca 2002). Although the final dosage of carbamazepine was in this case relatively low (400 mg/day), the blood level (2.4 µg/mL) was somewhat lower than would have been expected even so. This is likely because the phenytoin was chronically inducing the increased production of 3A4 and 2C9, so that the amount of 3A4 and 2C9 available for metabolizing carbamazepine was greater than would have otherwise been expected. The resulting greater metabolism of carbamazepine led to a lower than expected blood level.

Second, phenytoin is a substrate of 2C9 and 2C19 (Cadle et al. 1994; Mamiya et al. 1998), and carbamazepine is an inducer of 3A4, 1A2, and 2C9 (Miners and Birkett 1998; Parker et al. 1998; Spina et al. 1996). Thus, the addition of carbamazepine led to an increase in the amount of 2C9 that was available to metabolize the phenytoin. This resulted in more efficient metabolism of the phenytoin and a corresponding decline in the phenytoin blood level.

This combination of a lower than expected carbamazepine level and a decline in the phenytoin blood level led to a marked increase in the frequency of the patient's seizures. Although it is initially counterintuitive to imagine that the addition of a second anticonvulsant would lead to a loss of anticonvulsant efficacy, the reciprocal induction (or "coinduction") of these agents rendered the whole much less than the sum of its parts.

References

Cadle RM, Zenon GJ 3rd, Rodriguez-Barradas MC, et al: Fluconazole-induced symptomatic phenytoin toxicity. Ann Pharmacother 28(2):191–195, 1994

Chetty M, Miller R, Seymour MA: Phenytoin autoinduction. Ther Drug Monit 20(1):60–62, 1998

Mamiya K, Ieiri I, Shimamoto J, et al: The effects of genetic polymorphisms of CYP2C9 and CYP2C19 on phenytoin metabolism in Japanese adult patients with epilepsy: studies in stereoselective hydroxylation and population pharmacokinetics. Epilepsia 39(12):1317–1323, 1998

Miners JO, Birkett DJ: Cytochrome P4502C9: an enzyme of major importance in human drug metabolism. Br J Clin Pharmacol 45(6):525–538, 1998

Parker AC, Pritchard P, Preston T, et al: Induction of CYP1A2 activity by carbamazepine in children using the caffeine breath test. Br J Clin Pharmacol 45(2): 176–178, 1998

Spina E, Perucca E: Clinical significance of pharmacokinetic interactions between antiepileptic and psychotropic drugs. Epilepsia 43 (suppl 2):37–44, 2002

Spina E, Pisani F, Perucca E: Clinically significant pharmacokinetic drug interactions with carbamazepine: an update. Clin Pharmacokinet 31(3):198–214, 1996

Competitive Induction

A 69-year-old man with a seizure disorder, a history of deep venous thrombosis leading to pulmonary embolus, and polysubstance dependence (in full remission) had been maintained on phenytoin (Dilantin), 300 mg/day (blood level=10.7 μg/mL), and warfarin (Coumadin), 2.5 mg/day (International Normalized Ratio, or INR=2.1). He developed some sleep difficulties, but he did not want to take any sedative-hypnotic agents, for fear of triggering an addictive spiral of use. Trazodone (Desyrel) was not helpful. After some discussion with his internist, he was started on quetiapine (Seroquel), which was titrated to a dosage of 200 mg/day. This helped somewhat with his sleep. One day, he accidentally bumped his elbow on a door frame and was alarmed at the massive hematoma that formed at that area. He reported this event to his internist, who obtained a stat blood draw, which revealed an INR of 9.6 (Rogers et al. 1999).

Discussion

This is an example (I believe) of a substrate added to an inducer.

Quetiapine is a substrate primarily of 3A4 (DeVane and Nemeroff 2001), and phenytoin is an inducer of 3A4, 2C9, and 2C19 (Chetty et al. 1998; Spina and Perucca 2002). *S*-Warfarin (the primary active form) is primarily a substrate of 2C9 (Heimark et al. 1987; Linder and Valdes 1999), while *R*-warfarin is metabolized at both 1A2 and 3A4 (Lehmann 2000). Prior to the addition of the quetiapine, phenytoin was 1) inducing 2C9 to metabolize *S*-warfarin more efficiently than would otherwise have occurred, thus producing a lower INR than would have occurred at this dosage in the absence of phenytoin, and 2) inducing 3A4 to metabolize *R*-warfarin more efficiently than otherwise and thus lower the blood level of *R*-warfarin. Even though *S*-warfarin is by far the more active isomer of warfarin, there exists an equilibrium between these isomers, such that influences that alter the concentration of *R*-warfarin can secondarily, but significantly, alter the concentration of *S*-warfarin.

With the addition of quetiapine, some of the induced 3A4 that had been metabolizing the *R*-warfarin was now metabolizing quetiapine. This led to a net decrease in the metabolism of *R*-warfarin by 3A4, which led to an increase in the blood level of *R*-warfarin. This *R*-warfarin increase altered the equilibrium with *S*-warfarin, leading to a significant enough increase in the blood level of *S*-warfarin to significantly raise the patient's INR, even though the dosage of warfarin had not been changed.

This explanation is more conjectural than factual. However, 3A4 inhibitors, such as erythromycin and clarithromycin, do reliably increase warfarin blood levels, and it is believed that this occurs initially through increasing the levels of *R*-warfarin specifically (Oberg 1998). If this is truly the case, one might reasonably expect that a similar effect could be produced by decreasing inductive influences through the introduction of another substrate for the inducer in question. In this case, quetiapine seems to have lessened the effect of phenytoin's induction of *R*-warfarin's metabolism.

References

Chetty M, Miller R, Seymour MA: Phenytoin autoinduction. Ther Drug Monit 20(1):60–62, 1998

DeVane CL, Nemeroff CB: Clinical pharmacokinetics of quetiapine: an atypical antipsychotic. Clin Pharmacokinet 40(7):509–522, 2001

Heimark LD, Gibaldi M, Trager WF, et al: The mechanism of the warfarin-rifampin drug interaction. Clin Pharmacol Ther 42(4):388–394, 1987

Lehmann DE: Enzymatic shunting: resolving the acetaminophen-warfarin controversy. Pharmacotherapy 20(12):1464–1468, 2000

Linder MW, Valdes R Jr: Pharmacogenetics in the practice of laboratory medicine. Mol Diagn 4(4):365–379, 1999

Oberg KC: Delayed elevation of international normalized ratio with concurrent clarithromycin and warfarin therapy. Pharmacotherapy 18(2):386–391, 1998

Rogers T, deLeon J, Atcher D: Possible interaction between warfarin and quetiapine. J Clin Psychopharmacol 19(4):382–383, 1999

Spina E, Perucca E: Clinical significance of pharmacokinetic interactions between antiepileptic and psychotropic drugs. Epilepsia 43 (suppl 2):37–44, 2002

Chapter 6

Complex P450 Case Vignettes

The cases in this chapter represent the significant involvement of more than one P450 enzyme, or the presence of P450 and non-P450 influences in which the P450 influences are dominant. They are often more subtle and contain multiple interactive variables. These cases can even challenge the detection abilities of accomplished P450 veterans.

Too Much of a Good Thing (I)

A 66-year-old woman with previously diagnosed dysthymic disorder and significant coronary artery disease had been maintained on bupropion (Wellbutrin). Quinidine (Quinaglute) was one of her cardiac medications. Because her internist was both unfamiliar with and uncomfortable with bupropion, he discontinued the bupropion and started the patient on fluoxetine (Prozac), 20 mg/day. Over the next 1–2 weeks, the patient developed a progressive syndrome characterized by severe agitation, myoclonus, flushing, diarrhea, irritability, intermittent disorientation to place and time, and visual hallucinations of little men playing musical instruments at the foot of her bed without an auditory component (N.B. Sandson, self-report, October 1995).

Discussion

This case is a combined example of an inhibitor (quinidine) added to a substrate (fluoxetine) and a substrate (quinidine) added to an inhibitor (fluoxetine).

First, quinidine is one of the most potent 2D6 inhibitors in existence (von Moltke et al. 1994), and fluoxetine, a selective serotonin reuptake inhibitor, is a substrate of 2D6, among other P450 enzymes (Greenblatt et al. 1999; Ring et al. 2001). At 20 mg/day, the fluoxetine was being given at a dosage that would be appropriate for antidepressant action. However, the quinidine impaired the ability of 2D6 to make a significant contribution to the metabolism of fluoxetine, which led to an increase in the blood level of fluoxetine, even though the fluoxetine dosage had not been changed. This elevation in the fluoxetine blood level was sufficient to produce a central serotonin syndrome in this case. This very serious and potentially lethal syndrome is characterized by a systemic excess of serotonergic activity.

Second, quinidine is a 3A4 substrate (Koley et al. 1995), and fluoxetine (in concert with its active metabolite, norfluoxetine) is a strong 2D6 inhibitor and a moderate inhibitor of 3A4 (Greenblatt et al. 1999; Stevens and Wrighton 1993). Thus, the fluoxetine impaired the ability of 3A4 to efficiently metabolize the quinidine, likely resulting in an increase in the blood level of quinidine, even though the quinidine dosage had not been changed. The visual hallucinations in this case likely represent a state of quinidine toxicity superimposed on the aforementioned central serotonin syndrome. While this seems reasonable, unfortunately no quinidine levels were drawn to verify this conjecture.

References

Greenblatt DJ, von Moltke LL, Harmatz JS, et al: Human cytochromes and some newer antidepressants: kinetics, metabolism, and drug interactions. J Clin Psychopharmacol 19 (5, suppl 1):23S–35S, 1999

Koley AP, Buters JT, Robinson RC, et al: CO binding kinetics of human cytochrome P450 3A4: specific interaction of substrates with kinetically distinguishable conformers. J Biol Chem 270(10):5014–5018, 1995

Ring BJ, Eckstein JA, Gillespie JS, et al: Identification of the human cytochromes p450 responsible for in vitro formation of R- and S-norfluoxetine. J Pharmacol Exp Ther 297(3):1044–1050, 2001

Stevens JC, Wrighton SA: Interaction of the enantiomers of fluoxetine and norfluoxetine with human liver cytochromes P450. J Pharmacol Exp Ther 266(2): 964–971, 1993

von Moltke LL, Greenblatt DJ, Cotreau-Bibbo MM, et al: Inhibition of desipramine hydroxylation in vitro by serotonin-reuptake-inhibitor antidepressants, and by quinidine and ketoconazole: a model system to predict drug interactions in vivo. J Pharmacol Exp Ther 268(3):1278–1283, 1994

Too Much of a Good Thing (II)

A 49-year-old man was being treated with cimetidine (Tagamet), 300 mg qid, for reflux esophagitis, when he developed a cardiac arrhythmia. His cardiologist opted to start him on quinidine (Quinaglute), titrating the quinidine dose purely according to quinidine blood levels, until a therapeutic dosage (lower than a typical quinidine dosage, in this case) and level were achieved. This patient also had a remote history of major depressive disorder, and he suffered a recurrence of his depression following the news that his son had been placed in jail on charges of drug possession, breaking and entering, and aggravated assault. His internist noted his dysphoria, insomnia, and anhedonia, and he added paroxetine (Paxil), 20 mg/day, to the patient's regimen. Within 5 days, the patient was experiencing a full central serotonin syndrome, with flushing, delirium, fever, myoclonus, and vomiting. He was promptly hospitalized, his cimetidine and paroxetine were discontinued, and with antipyretics, muscle relaxants, and liberal quantities of lorazepam (Ativan), he made a full recovery in another 5 days (Pies 2002).

Discussion

This is an example (actually two examples) of a substrate added to an inhibitor.

Cimetidine is a potent inhibitor of 2D6, 3A4, and 1A2 (Martinez et al. 1999). Quinidine is both a substrate of 3A4 and an extremely potent 2D6 inhibitor (Koley et al. 1995; von Moltke et al. 1994). Paroxetine is metabolized principally by 2D6 (Paxil [package insert] 2001), and it is a potent 2D6 inhibitor (von Moltke et al. 1995). When the paroxetine was added to the cimetidine and quinidine, 2D6 was severely impaired in its ability to metabolize the paroxetine, which led to a significant increase in the blood level of paroxetine. This escalation of the paroxetine blood level caused excessive inhibition of serotonin reuptake, leading to the central serotonin syndrome.

It is worth noting that there was not a superimposed element of quinidine toxicity as there was in the previous case ("Too Much of a Good Thing [I]"), when fluoxetine was added to quinidine. The reasons are twofold. First, the shrewd decision was made to titrate the quinidine dose according to blood levels, which automatically took cimetidine's 3A4 inhibition of quinidine's metabolism into account, rather than to titrate quinidine according to preset dosing guidelines—a dosing strategy that would likely have led to a state of quinidine toxicity. Second, fluoxetine is a moderate inhibitor of 3A4, whereas paroxetine's 3A4 inhibition is much weaker. Thus, fluoxetine was more able to impair the metabolism of quinidine at 3A4, and thus generate increased quinidine blood levels, than was paroxetine.

References

Koley AP, Buters JT, Robinson RC, et al: CO binding kinetics of human cytochrome P450 3A4: specific interaction of substrates with kinetically distinguishable conformers. J Biol Chem 270(10):5014–5018, 1995

Martinez C, Albet C, Agundez JA, et al: Comparative in vitro and in vivo inhibition of cytochrome P450 CYP1A2, CYP2D6, and CYP3A by H2-receptor antagonists. Clin Pharmacol Ther 65(4):369–376, 1999

Paxil (package insert). Research Triangle Park, NC, GlaxoSmithKline, 2001

Pies R: Cytochromes and beyond: drug interactions in psychiatry. Psychiatric Times, May 2002, pp 48–51

von Moltke LL, Greenblatt DJ, Cotreau-Bibbo MM, et al: Inhibition of desipramine hydroxylation in vitro by serotonin-reuptake-inhibitor antidepressants, and by quinidine and ketoconazole: a model system to predict drug interactions in vivo. J Pharmacol Exp Ther 268(3):1278–1283, 1994

von Moltke LL, Greenblatt DJ, Court MH, et al: Inhibition of alprazolam and desipramine hydroxylation in vitro by paroxetine and fluvoxamine: comparison with other selective serotonin reuptake inhibitor antidepressants. J Clin Psychopharmacol 15(2):125–131, 1995

Can't Sit Still

A 23-year-old man with schizophrenia was hospitalized for an exacerbation of psychotic symptoms. On admission, he was restarted on the haloperidol (Haldol) that he had discontinued 1 month prior to admission. The haloperidol was titrated to a dosage of 10 mg/day. After 10 days at this dosage, his psychotic symptoms had significantly remitted, but an array of depressive symptoms (dysphoria, anhedonia, terminal insomnia, poor appetite, and poor energy) began to emerge. Fluvoxamine (Luvox) was added and titrated over a 2-week period to a dose of 100 mg bid to address these depressive symptoms. Five days later, the patient began to experience an unpleasant sensation of severe restlessness, and he had great difficulty remaining seated for more than 2 minutes at a time.

Discussion

This is an example of an inhibitor added to a substrate.

Fluvoxamine is a potent inhibitor of 1A2 and a moderate inhibitor of 3A4 (Christensen et al. 2002; von Moltke et al. 1995). Haloperidol is metabolized by 3A4, 2D6, 1A2, and phase II glucuronidation (Desai et al. 2001; Kudo and Ishizaki 1999). At a dosage of 10 mg/day, haloperidol was effective in treating the patient's psychotic symptoms and no side effects were evident. However, with the introduction of fluvoxamine, which significantly inhibited haloperidol's metabolism at 1A2 and 3A4, the serum concentration of haloperidol rose markedly. There are reports that the addition of fluvoxamine to haloperidol has led to three- to sixfold increases in haloperidol blood levels (Brosen 1995). In this case, the interaction led to the emergence of akathisia, a form of antipsychotic-induced extrapyramidal symptoms.

References

Brosen K: Drug interactions and the cytochrome P450 system: the role of cytochrome P450 1A2. Clin Pharmacokinet 29 (suppl 1):20–25, 1995

Christensen M, Tybring G, Mihara K, et al: Low daily 10-mg and 20-mg doses of fluvoxamine inhibit the metabolism of both caffeine (cytochrome P4501A2) and omeprazole (cytochrome P4502C19). Clin Pharmacol Ther 71:141–152, 2002

Desai HD, Seabolt J, Jann MW: Smoking in patients receiving psychotropic medications: a pharmacokinetic perspective. CNS Drugs 15(6):469–494, 2001

Kudo S, Ishizaki T: Pharmacokinetics of haloperidol: an update. Clin Pharmacokinet 37(6):435–456, 1999

von Moltke LL, Greenblatt DJ, Court MH, et al: Inhibition of alprazolam and desipramine hydroxylation in vitro by paroxetine and fluvoxamine: comparison with other selective serotonin reuptake inhibitor antidepressants. J Clin Psychopharmacol 15(2):125–131, 1995

Less Is More

A 32-year-old man with schizoaffective disorder, bipolar type, had been noncompliant with his olanzapine (Zyprexa), 20 mg qhs, which led to a psychotic decompensation with manic features and subsequent hospitalization. His inpatient psychiatrist decided that he should begin receiving a decanoate preparation of haloperidol (Haldol) to address his recurrent medication noncompliance. In preparation for this, he was started on oral haloperidol, 10 mg/day. Also, since his current presentation was characterized by more violence than usual, the psychiatrist added carbamazepine (Tegretol), titrating to a dosage of 800 mg/day (blood level= 9.7 μg/mL). After 2 weeks at this dosage of carbamazepine, the patient was slightly calmer, but his paranoid delusions and auditory hallucinations persisted with unabated strength. The haloperidol was further increased to 20 mg/day in an effort to treat these psychotic symptoms. Ten days later, the psychotic symptoms began to diminish, but the psychiatrist discovered that the patient's white blood count had dropped from 6,100 cells/μL to 2,100 cells/μL. The carbamazepine was abruptly discontinued. One week later, with the patient taking the same dosage of haloperidol, the patient's psychotic symptoms continued to improve. At about the same time, however, he began to experience cogwheel rigidity and mild akathisia. The psychiatrist added benztropine (Cogentin) for the cogwheeling and strongly considered adding propranolol (Inderal) for the akathisia, but following a conversation with the hospital pharmacist, he instead chose to decrease the haloperidol back to 10 mg/day. One week later, the patient's psychotic symptoms had almost completely remitted, as had all of his side effects, including the akathisia.

Discussion

This is an example of reversal of induction.

Haloperidol is metabolized by 3A4, 2D6, 1A2, and phase II glucuronidation (Desai et al. 2001; Kudo and Ishizaki 1999), and carbamazepine is an inducer of 3A4, 1A2, and phase II glucuronidation (Hachad et al. 2002; Lucas et al. 1998; Parker et al. 1998; Spina et al. 1996). The carbamazepine increased the amounts of 3A4, 1A2, and possibly pertinent phase II glucuronidation enzymes that were available to metabolize the haloperidol, leading to a haloperidol blood level that was potentially subtherapeutic despite the use of fairly high doses (Hesslinger et al. 1999). When the carbamazepine was discontinued, the "extra" 3A4, 1A2, and possibly pertinent glucuronidation enzymes that were produced by carbamazepine's induction "died off," and metabolically available levels of 3A4, 1A2, and glucuronidation enzymes returned to their (lower) precarbamazepine baselines. This led to a decrease in the metabolism of haloperidol and a subsequent increase in the haloperidol blood level. In this case, the resulting blood level was now sufficient to generate both an antipsychotic response, and parkinsonism and akathisia. A decrease in the haloperidol dosage from 20 mg/day to 10 mg/day, along with a little benztropine, allowed for the remission of these side effects with a continuation of the therapeutic response.

References

Desai HD, Seabolt J, Jann MW: Smoking in patients receiving psychotropic medications: a pharmacokinetic perspective. CNS Drugs 15(6):469–494, 2001

Hachad H, Ragueneau-Majlessi I, Levy RH: New antiepileptic drugs: review on drug interactions. Ther Drug Monit 24(1):91–103, 2002

Hesslinger B, Normann C, Langosch JM, et al: Effects of carbamazepine and valproate on haloperidol plasma levels and on psychopathologic outcome in schizophrenic patients. J Clin Psychopharmacol 19(4):310–315, 1999

Kudo S, Ishizaki T: Pharmacokinetics of haloperidol: an update. Clin Pharmacokinet 37(6):435–56, 1999

Lucas RA, Gilfillan DJ, Bergstrom RF: A pharmacokinetic interaction between carbamazepine and olanzapine: observations on possible mechanism. Eur J Clin Pharmacol 54(8):639–643, 1998

Parker AC, Pritchard P, Preston T, et al: Induction of CYP1A2 activity by carbamazepine in children using the caffeine breath test. Br J Clin Pharmacol 45(2):176–178, 1998

Spina E, Pisani F, Perucca E: Clinically significant pharmacokinetic drug interactions with carbamazepine: an update. Clin Pharmacokinet 31(3):198–214, 1996

Structural Similarities and Cardiac Calamities

A 55-year-old woman with a history of major depressive disorder, in full remission, fell from a ladder while mending a shingle on her house and fractured her hip. She was hospitalized for a sustained period, and she experienced a recurrence of her depression as a consequence of her current disability. Her previous depressive episodes had responded to amitriptyline (Elavil), 200 mg/day, although she had not taken this medication for more than 8 years. After conferring with her previous psychiatrist, the surgeon started the patient on amitriptyline, titrating to a dose of 150 mg qhs. She was also receiving hydromorphone to treat her postsurgical pain, and this was providing adequate analgesia. She was, however, complaining about muscle stiffness and cramping, so the surgeon added cyclobenzaprine (Flexeril), 10 mg tid. Over the next week, the patient became increasingly confused, with alternating periods of somnolence and increasingly severe agitation and irritability. She eventually became frankly combative and required restraints. At this point, her surgeon ordered two successive doses of haloperidol (Haldol), 5 mg im, roughly 3 hours apart, in order to help calm her. Unfortunately, soon thereafter she experienced a cardiac arrest and expired despite efforts at resuscitation (R. Love, personal communication, August 2002).

Discussion

This is an example of multiple "inhibitor added to substrate" and "substrate added to inhibitor" combinations, synergizing with additive pharmacodynamic effects on cardiac conduction.

Amitriptyline, cyclobenzaprine, and haloperidol are all partially metabolized at 1A2 (Desai et al. 2001; Venkatakrishnan et al. 1998; Wang et al. 1996). Thus, when the cyclobenzaprine was added to the amitriptyline, each agent likely acted as both a competitive inhibitor slowing the metabolism of the other agent at 1A2, and a substrate whose metabolism at 1A2 was being impaired by the introduction of the other agent acting as a competitive inhibitor. With the addition of the haloperidol to the combination of amitriptyline and cyclobenzaprine, this became a three-way reciprocal interaction, the net effect being some degree of elevation of the blood levels of all three agents. (Although the haloperidol had only just been introduced, its intramuscular administration may well have enabled it to meaningfully participate in this three-way interaction.) Unfortunately, no blood levels were drawn to verify that such an interaction was taking place, and therefore the pharmacokinetic component of this drug interaction remains more conjectural than factual.

The pharmacodynamic synergy between these agents is more straightforward, and it is probably the more important contributor to this adverse event. Cyclobenzaprine is very closely related in structure to the tricyclic antidepressants (TCAs), and thus it shares their quinidine-like effects on cardiac conduction. Haloperidol also has a recognized ability to prolong the QTc interval (Goodnick et al. 2002). When these three agents were coadministered, their likely mutual inhibition of one another's metabolism, resulting in blood level elevations of all three agents, likely augmented their shared effects on cardiac conduction to produce this fatal outcome.

References

Desai HD, Seabolt J, Jann MW: Smoking in patients receiving psychotropic medications: a pharmacokinetic perspective. CNS Drugs 15(6):469–494, 2001

Goodnick PJ, Jerry J, Parra F: Psychotropic drugs and the ECG: focus on the QTc interval. Expert Opin Pharmacother 3(5):479–498, 2002

Venkatakrishnan K, Greenblatt DJ, von Moltke LL, et al: Five distinct human cytochromes mediate amitriptyline N-demethylation in vitro: dominance of CYP 2C19 and 3A4. J Clin Pharmacol 38(2):112–121, 1998

Wang RW, Liu L, Cheng H: Identification of human liver cytochrome P450 isoforms involved in the in vitro metabolism of cyclobenzaprine. Drug Metab Dispos 24(7):786–791, 1996

Dangerous Disinhibition

A 35-year-old woman with severe, recurrent major depression and chronic gastroesophageal reflux disease (GERD) had been reasonably stable on the following regimen: fluoxetine (Prozac), 60 mg/day; doxepin (Sinequan), 25 mg/day (blood level=generally around 250 ng/mL); and cimetidine (Tagamet), 300 mg qid. Her depression had not previously responded to antidepressant monotherapy with fluoxetine, 60 mg/day. Her GERD began to worsen, and she experienced more heartburn in the evenings. She consulted with her gastroenterologist, who decided to discontinue the cimetidine and instead have her start taking pantoprazole (Protonix). Within 5 weeks, she became desperately depressed and required psychiatric hospitalization following an overdose attempt. A doxepin blood level performed on admission was only 104 ng/mL (D. Benedek, personal communication, May 2002).

Discussion

This is an example of reversal of inhibition.

Doxepin is a tertiary-amine tricyclic antidepressant (TCA) whose metabolism depends most on the intact functioning of 2C19, 3A4, 2D6, and 1A2 in a manner similar to that of imipramine (Yang et al. 1999). Cimetidine is a potent pan-inhibitor of all major P450 enzymes except for 2E1 (Martinez et al. 1999; Pies 2002), and fluoxetine (in concert with its active metabolite, norfluoxetine) is a strong 2D6 inhibitor and a moderate 3A4 inhibitor (Greenblatt et al. 1999; Stevens and Wrighton 1993). The discontinuation of the cimetidine led to decreased inhibition of the ability of 2D6, 3A4, 2C19, and 1A2 to metabolize the doxepin, with a resulting decrease in the doxepin blood level, even though the dosage had not been changed. The continuing 2D6 and 3A4 inhibition provided by fluoxetine still yielded a doxepin blood level (104 ng/mL) that was higher than what would have been obtained had doxepin been given as antidepressant monotherapy at the very low dosage of 25 mg/day (150–200 mg/day is more typical), but this blood level was not sufficient to prevent a relapse of depression. Thus, the patient's therapeutic response to doxepin, at this dosage, was dependent on the presence of robust inhibitors of its metabolism. Removal of one of those inhibitors led to a decline in the doxepin blood level and a loss of antidepressant efficacy.

References

Greenblatt DJ, von Moltke LL, Harmatz JS, et al: Human cytochromes and some newer antidepressants: kinetics, metabolism, and drug interactions. J Clin Psychopharmacol 19 (5, suppl 1):23S–35S, 1999

Martinez C, Albet C, Agundez JA, et al: Comparative in vitro and in vivo inhibition of cytochrome P450 CYP1A2, CYP2D6, and CYP3A by H2-receptor antagonists. Clin Pharmacol Ther 65(4):369–376, 1999

Pies R: Cytochromes and beyond: drug interactions in psychiatry. Psychiatric Times, May 2002, pp 48–51

Stevens JC, Wrighton SA: Interaction of the enantiomers of fluoxetine and norfluoxetine with human liver cytochromes P450. J Pharmacol Exp Ther 266(2): 964–971, 1993

Yang TJ, Krausz KW, Sai Y, et al: Eight inhibitory monoclonal antibodies define the role of individual P450s in human liver microsomal diazepam, 7-ethoxycoumarin, and imipramine metabolism. Drug Metab Dispos 27(1):102–109, 1999

Rumination

A 27-year-old woman with a history of polysubstance abuse, in remission, and migraine headaches had been receiving imipramine (Tofranil), 150 mg/day (imipramine+desipramine blood level=roughly 250 ng/mL) for the past 3 years, with generally good results. However, over the past 6 months her migraines had become more frequent and severe. This had led her to become very tentative about taking trips, making evening plans, or engaging in any behavior that might make it difficult for her to obtain emergent relief from her migraine pain. The patient and her neurologist had been reluctant to prescribe any controlled substances, given the patient's history of addiction. During her most recent visit to her neurologist, he was impressed by the strength of the patient's anxious ruminations about her migraine pain, and he wondered if her secondary anxiety might not be compounding her pain. The neurologist had heard that 2D6 inhibitors could elevate tricyclic antidepressant (TCA) blood levels, so he wanted to avoid any such medications. He had also heard that fluvoxamine (Luvox) was especially helpful for obsessive-compulsive disorder (OCD). Given her persistent ruminations, he thought fluvoxamine would be a good choice. He had the patient begin taking fluvoxamine, at a dosage of 50 mg/day, and titrated the dosage to 150 mg/day over 10 days. After 2 days on this dosage, the patient reported light-headedness, blurry vision, constipation, sedation, and palpitations. Fearing the worst, the neurologist wisely advised her to report to the nearest ER, where her imipramine+desipramine blood level was found to be 973 ng/mL. Although her electrocardiogram revealed only sinus tachycardia, she was nonetheless admitted to a telemetry unit for observation.

Discussion

This is an example of an inhibitor added to a substrate.

Imipramine is a tertiary-amine TCA whose metabolism depends most on the intact functioning of 2C19, 3A4, 2D6, and 1A2 (Yang et al. 1999). Fluvoxamine is a strong inhibitor of 1A2, 2C9, and 2C19 and a moderate inhibitor of 3A4 (Christensen et al. 2002; Niemi et al. 2001; von Moltke et al. 1995). Thus, the addition of fluvoxamine significantly impaired the ability of these enzymes to make an effective contribution to the metabolism of imipramine. Although 2D6 was able to contribute to imipramine's metabolism, the inhibition of all the other pertinent P450 enzymes clearly overwhelmed 2D6's capabilities, and the blood level of imipramine + desipramine (imipramine's primary metabolite) rose nonetheless. Studies have demonstrated that the addition of fluvoxamine to imipramine can cause a three- to sixfold increase in blood levels (Brosen 1995).

Although a breakdown of the individual imipramine+desipramine blood levels was not available from this ER, it would stand to reason that the imipramine portion of the sum would be greatly elevated, while the desipramine portion might be unaffected or minimally elevated. This would be due to the inhibition of 3A4 and 2C19 impairing the conversion of imipramine into desipramine via demethylation, while the functioning of the principal enzyme responsible for desipramine's metabolism (2D6) continued to function virtually at its baseline level of activity. This pattern would be analogous to those seen in the cases of clomipramine added to fluvoxamine (see "Symmetry," p. 184) and nefazodone added to amitriptyline (see "Vigilance Always," p. 192).

References

Brosen K: Drug interactions and the cytochrome P450 system: the role of cytochrome P450 1A2. Clin Pharmacokinet 29 (suppl 1):20–25, 1995

Christensen M, Tybring G, Mihara K, et al: Low daily 10-mg and 20-mg doses of fluvoxamine inhibit the metabolism of both caffeine (cytochrome P4501A2) and omeprazole (cytochrome P4502C19). Clin Pharmacol Ther 71:141–152, 2002

Niemi M, Backman JT, Neuvonen M, et al: Effects of fluconazole and fluvoxamine on the pharmacokinetics and pharmacodynamics of glimepiride. Clin Pharmacol Ther 69(4):194–200, 2001

von Moltke LL, Greenblatt DJ, Court MH, et al: Inhibition of alprazolam and desipramine hydroxylation in vitro by paroxetine and fluvoxamine: comparison with other selective serotonin reuptake inhibitor antidepressants. J Clin Psychopharmacol 15(2):125–131, 1995

Yang TJ, Krausz KW, Sai Y, et al: Eight inhibitory monoclonal antibodies define the role of individual P450s in human liver microsomal diazepam, 7-ethoxycoumarin, and imipramine metabolism. Drug Metab Dispos 27(1):102–109, 1999

Enuresis (II)

A 10-year-old boy with persistent nocturnal enuresis, which had not responded to behavioral interventions, responded to a trial of imipramine (Tofranil), 50 mg/day (no blood levels were obtained). His parents were awaiting a consultation with a pediatric urologist when the patient experienced a seizure. He was eventually placed on carbamazepine (Tegretol), 600 mg/day (blood level=6.8 μg/mL), in addition to his imipramine, to address his new-onset seizure disorder. Within 2 weeks, the patient was again consistently enuretic at night.

Discussion

This is an example of an inducer added to a substrate.

Imipramine is a tertiary-amine tricyclic antidepressant (TCA) whose metabolism depends most on the intact functioning of 2C19, 3A4, 2D6, and 1A2 (Yang et al. 1999). Carbamazepine is an inducer of 3A4 and 1A2 (Parker et al. 1998; Spina et al. 1996). Thus, the addition of carbamazepine led to increased amounts of 3A4 and 1A2 that were available to metabolize the imipramine at a more rapid rate. This led to a decrease in the blood level of imipramine+desipramine (imipramine's primary metabolite) (Brown et al. 1990). Since the efficacy of imipramine for nocturnal enuresis presumably relies on its dose-dependent (and therefore blood level–dependent) anticholinergic properties, this decrease in the imipramine+desipramine blood level led to a loss of efficacy in treating this condition. Therapeutic efficacy could be recaptured by increasing the imipramine dosage to compensate for carbamazepine's induction of imipramine's metabolism, but care would have to be taken to reduce the imipramine dosage if the carbamazepine were ever removed. TCA levels should naturally be closely monitored, especially in children and adolescents, whose mass and metabolism are constantly in flux.

References

Brown CS, Wells BG, Cold JA, et al: Possible influence of carbamazepine on plasma imipramine concentrations in children with attention deficit hyperactivity disorder. J Clin Psychopharmacol 10(5):359–362, 1990

Parker AC, Pritchard P, Preston T, et al: Induction of CYP1A2 activity by carbamazepine in children using the caffeine breath test. Br J Clin Pharmacol 45(2):176–178, 1998

Spina E, Pisani F, Perucca E: Clinically significant pharmacokinetic drug interactions with carbamazepine: an update. Clin Pharmacokinet 31(3):198–214, 1996

Yang TJ, Krausz KW, Sai Y, et al: Eight inhibitory monoclonal antibodies define the role of individual P450s in human liver microsomal diazepam, 7-ethoxycoumarin, and imipramine metabolism. Drug Metab Dispos 27(1):102–109, 1999

Hematuria

A 72-year-old woman with type II diabetes mellitus and atrial fibrillation was being well maintained on warfarin (Coumadin), 5 mg/day (International Normalized Ratio, or INR=2.9); metoprolol (Lopressor), 25 mg/day; and amitriptyline (Elavil), 50 mg qhs, for neuropathic pain. The patient scheduled an appointment with her internist to discuss her persistent anergy and insomnia. As they spoke, the internist discerned that the patient was struggling with the recent death of her husband and was frankly depressed. He chose to start the patient on fluoxetine (Prozac), 20 mg/day. Ten days later, the patient complained about dizziness, dry mouth, and inability to void. She contacted her internist, who advised her to call 911 and have an ambulance transport her to the nearest ER. Once there, a bladder catheterization yielded 2 liters of dark urine. Her INR was found to be 17.3. No amitriptyline/nortriptyline levels were obtained (J.R. Oesterheld, personal communication, July 2002).

Discussion

This is an example of an inhibitor added to two substrates, whose effects synergized to produce the complication described above.

First, warfarin's metabolism is exceedingly complex, but it mostly occurs at 2C9 for the more active *S*-warfarin isomer (Heimark et al. 1987; Linder and Valdes 1999). The less active *R*-warfarin isomer is metabolized primarily at 1A2 (Lehmann 2000). Fluoxetine (in concert with its active metabolite, norfluoxetine) is a strong 2D6 inhibitor and a moderate inhibitor of 2C9, 2C19, and 3A4 (Greenblatt et al. 1999; Stevens and Wrighton 1993). Thus, the addition of the fluoxetine significantly impaired the ability of 2C9 to efficiently metabolize the warfarin, which led to an increase in the warfarin blood level, even though the warfarin dosage had not been changed. This increase in the blood level of warfarin drastically increased the magnitude of warfarin's anticoagulant effect.

Second, amitriptyline is a tertiary-amine TCA whose metabolism depends most on the intact functioning of 2C19, 3A4, and 2D6, with 1A2 serving as a secondary enzyme (Venkatakrishnan et al. 1998). The fluoxetine significantly impaired the ability of 2D6, 3A4, and 2C19 to contribute to the metabolism of amitriptyline, which led to an increase in the blood level of amitriptyline+nortriptyline. Even though this was unconfirmed by an actual blood level, a process along these lines is reasonably inferable, as evidenced by the increase in the anticholinergic effects of amitriptyline, leading to inability to void.

The combination of 1) fluoxetine's elevation of the amitriptyline level (leading to anticholinergic-induced inability to void and subsequent bladder distension) and 2) fluoxetine's elevation of the warfarin level (leading to a severely hypocoagulable state) caused spontaneous bleeding within the patient's bladder. This explains both the quantity and the color of the urine drained by catheterization of her bladder.

References

Greenblatt DJ, von Moltke LL, Harmatz JS, et al: Human cytochromes and some newer antidepressants: kinetics, metabolism, and drug interactions. J Clin Psychopharmacol 19 (5, suppl 1):23S–35S, 1999

Heimark LD, Gibaldi M, Trager WF, et al: The mechanism of the warfarin-rifampin drug interaction. Clin Pharmacol Ther 42(4):388–394, 1987

Lehmann DE: Enzymatic shunting: resolving the acetaminophen-warfarin controversy. Pharmacotherapy 20(12):1464–1468, 2000

Linder MW, Valdes R Jr: Pharmacogenetics in the practice of laboratory medicine. Mol Diagn 4(4):365–379, 1999

Stevens JC, Wrighton SA: Interaction of the enantiomers of fluoxetine and norfluoxetine with human liver cytochromes P450. J Pharmacol Exp Ther 266(2): 964–971, 1993

Venkatakrishnan K, Greenblatt DJ, von Moltke LL, et al: Five distinct human cytochromes mediate amitriptyline N-demethylation in vitro: dominance of CYP 2C19 and 3A4. J Clin Pharmacol 38(2):112–121, 1998

Antiemetic Arrhythmia

A 61-year-old woman with recurrent major depression was admitted to a surgical service for an orthopedic procedure. Her regular medications included fluoxetine (Prozac), 20 mg/day, and cyclobenzaprine (Flexeril),

10 mg tid. Her preoperative electrocardiogram revealed a QTc interval of 495 msec. The patient reported that she had become nauseated the last time she was given a general anesthetic, so she received preoperative droperidol (Inapsine), 5 mg iv. During the surgical procedure, the patient developed a torsades de pointes arrhythmia that progressed into ventricular fibrillation. She was successfully restored to normal sinus rhythm, and her fluoxetine and cyclobenzaprine were discontinued for the time being. Three days later, her QTc interval had decreased to 436 msec (Michalets et al. 1998).

Discussion

This is an example of multiple substrate-inhibitor interactions synergizing with pharmacodynamic interactions between agents.

Cyclobenzaprine is a substrate of both 1A2 and 3A4 (Wang et al. 1996), although its tricyclic structure suggests that it may also rely on 2C9/19 and/or 2D6 for some of its metabolism. Fluoxetine (in concert with its active metabolite, norfluoxetine) is a strong 2D6 inhibitor and a moderate inhibitor of 3A4, 2C9, and 2C19 (Greenblatt et al. 1999; Stevens and Wrighton 1993). Thus, the presence of fluoxetine significantly impaired the ability of 3A4 (and possibly 2D6, 2C9, and 2C19 as well) to contribute to the metabolism of cyclobenzaprine, leading to a greater blood level of cyclobenzaprine than would have existed at this dosage had the fluoxetine not been present. Droperidol is a butyrophenone neuroleptic, structurally analogous to haloperidol (Haldol). The metabolism of droperidol is not well understood, but its similarity to haloperidol suggests that it may be metabolized by 2D6, 3A4, 1A2, and glucuronidation. If that is the case, fluoxetine would have impaired the ability of 2D6 and 3A4 to metabolize the droperidol, leading to a greater than expected blood level of droperidol following the administration of the intravenous droperidol. However, beyond the inhibition of cyclobenzaprine's metabolism at 3A4, these other pharmacokinetic considerations are conjectural.

Cyclobenzaprine (because of its close structural relationship to the tricyclic antidepressants [TCAs]) and droperidol both have the ability to prolong the QTc interval (Goodnick et al. 2002). Given that this patient had a quite long preoperative QTc interval, the administration of droperidol under these circumstances almost certainly led to additional prolongation of the QTc interval, which led to her cardiac arrhythmias. When her fluoxetine and cyclobenzaprine were discontinued, it was discovered that her preoperative QTc interval (495 msec) was significantly higher than her true baseline QTc interval (436 msec). This was likely due to the influence of cyclobenzaprine, whose metabolism was probably being slowed and whose blood level was probably being increased by fluoxetine, leading to greater QTc prolongation than would likely have occurred had the fluoxetine not been present. The pharmacokinetic interactions likely maximized the ability of these agents to exert their pharmacodynamic influence on this patient's cardiac conduction.

References

Goodnick PJ, Jerry J, Parra F: Psychotropic drugs and the ECG: focus on the QTc interval. Expert Opin Pharmacother 3(5):479–498, 2002

Greenblatt DJ, von Moltke LL, Harmatz JS, et al: Human cytochromes and some newer antidepressants: kinetics, metabolism, and drug interactions. J Clin Psychopharmacol 19 (5, suppl 1):23S–35S, 1999

Michalets EL, Smith LK, Van Tassel ED: Torsade de pointes resulting from the addition of droperidol to an existing cytochrome P450 drug interaction. Ann Pharmacother 32(7–8):761–765, 1998

Stevens JC, Wrighton SA: Interaction of the enantiomers of fluoxetine and norfluoxetine with human liver cytochromes P450. J Pharmacol Exp Ther 266(2):964–971, 1993

Wang RW, Liu L, Cheng H: Identification of human liver cytochrome P450 isoforms involved in the in vitro metabolism of cyclobenzaprine. Drug Metab Dispos 24(7):786–791, 1996

Gabriel

A 29-year-old man with chronic paranoid schizophrenia was being treated with haloperidol (Haldol), 10 mg/day, and benztropine (Cogentin), 2 mg/day. After close contact with a peer who had been diagnosed with tuberculosis, the patient also began to display chronic cough, malaise, and fevers. He reported to a local ER and voiced his concerns. He was diagnosed with tuberculosis and was started on, among other medications, rifampin (in the form of Rifadin). One month later, he took a cab to the local airport and proclaimed himself to be the angel Gabriel, sent by God to save the world from the hordes of the Martian Queen. He loudly and repeatedly accosted passersby until airport security eventually detained him. He was taken to the local ER, where he insisted that he had been compliant with his haloperidol. He was admitted to an inpatient psychiatry unit, where his haloperidol level was found to be less than 2 ng/mL. He was eventually stabilized with a haloperidol dosage of 25 mg/day, which generated a blood level of 7.3 ng/mL.

Discussion

This is an example of an inducer added to a substrate.

Haloperidol is metabolized by 3A4, 2D6, 1A2, and phase II glucuronidation (Desai et al. 2001; Kudo and Ishizaki 1999), while rifampin is a pan-inducer of 1A2, 3A4, 2C9, 2C19, and glucuronidation (Ebert et al. 2000; Heimark et al. 1987; Kay et al. 1985; Strayhorn et al. 1997; Wietholtz et al. 1995; Zhou et al. 1990; Zilly et al. 1977). Thus, the addition of rifampin led to the increased production of 3A4, 1A2, and glucuronidation enzymes, which were then able to more efficiently metabolize the haloperidol. This led to a decrease in the haloperidol blood level (DeVane and Nemeroff 2002) into the subtherapeutic range, which culminated in a recurrence of grandiose delusions and auditory hallucinations (instructions from God). The inpatient psychiatrist was able to eventually compensate for rifampin's induction of 3A4, 1A2, and glucuronidation by increasing the haloperidol dosage to 25 mg/day, which restored therapeutic efficacy without producing any new side effects.

References

Desai HD, Seabolt J, Jann MW: Smoking in patients receiving psychotropic medications: a pharmacokinetic perspective. CNS Drugs 15(6):469–494, 2001

DeVane CL, Nemeroff CB: 2002 guide to psychotropic drug interactions. Primary Psychiatry 9(3):28–57, 2002

Ebert U, Thong NQ, Oertel R, et al: Effects of rifampin and cimetidine on pharmacokinetics and pharmacodynamics of lamotrigine in healthy subjects. Eur J Clin Pharmacol 56(4):299–304, 2000

Heimark LD, Gibaldi M, Trager WF, et al: The mechanism of the warfarin-rifampin drug interaction. Clin Pharmacol Ther 42(4):388–394, 1987

Kay L, Kampmann JP, Svendsen TL, et al: Influence of rifampicin and isoniazid on the kinetics of phenytoin. Br J Clin Pharmacol 20(4):323–326, 1985

Kudo S, Ishizaki T: Pharmacokinetics of haloperidol: an update. Clin Pharmacokinet 37(6):435–456, 1999

Strayhorn VA, Baciewicz AM, Self TH: Update on rifampin drug interactions, III. Arch Intern Med 157(21):2453–2458, 1997

Wietholtz H, Zysset T, Marschall HU, et al: The influence of rifampin treatment on caffeine clearance in healthy man. J Hepatol 22(1):78–81, 1995

Zhou HH, Anthony LB, Wood AJ, et al: Induction of polymorphic 4′ hydroxylation of S-mephenytoin by rifampicin. Br J Clin Pharmacol 30(3):471–475, 1990

Zilly W, Breimer DD, Richter E: Stimulation of drug metabolism by rifampicin in patients with cirrhosis or cholestasis measured by increased hexobarbital and tolbutamide clearance. Eur J Clin Pharmacol 11(4):287–293, 1977

Symmetry

An 11-year-old boy was diagnosed with comorbid major depressive disorder and obsessive-compulsive disorder (OCD). His intrusive thoughts and rituals around preserving symmetry in his surroundings had significantly derailed his schoolwork and his ability to maintain normal peer relationships. The patient was initially tried on fluoxetine (Prozac), 20 mg/day, but he developed tinnitus while taking this medication. The fluoxetine was discontinued, and he was then started on fluvoxamine (Luvox). The dosage of this medication was initially titrated up to 150 mg/day, with mild to moderate OCD symptom reduction, but the patient developed a severe apathy syndrome at this dosage. The fluvoxamine dosage was decreased to 100 mg/day, with a remission of the apathy but also a loss of efficacy against the OCD symptoms. The psychiatrist decided to add clomipramine (Anafranil), initially at a dose of 25 mg qhs. After 10 days, blood levels were as follows: clomipramine level=130 ng/mL (optimal range for this lab was 70–200 ng/mL); desmethylclomipramine (D-CMI) level=41 ng/mL (range was 150–300 ng/mL); total=171 ng/mL (range was 220–500 ng/mL). The clomipramine dose was then increased to 50 mg qhs, yielding the following levels: clomipramine=352 ng/mL; D-CMI=52 ng/mL; total=404 ng/mL. Even though the clomipramine+D-CMI total was within the desired range, the clinician was concerned (erroneously) about the high clomipramine level leading to cardiac toxicity. He therefore decided to discontinue the fluvoxamine and started the patient on venlafaxine (Effexor). After the venlafaxine dosage was titrated to 112.5 mg/day, the patient's blood levels were as follows: clomipramine=24 ng/mL, D-CMI=103 ng/mL, and total=127 ng/mL. Curiously, the patient began to enjoy a positive clinical response, with remission of both depressive and OCD symptoms, without the need for further dose changes of either venlafaxine or clomipramine (C. Lachner, personal communication, August 2002).

Discussion

This is an example of a substrate added to an inhibitor.

Clomipramine is a tertiary-amine tricyclic antidepressant (TCA) whose metabolism depends most on the intact functioning of 2C19, 3A4, and 2D6, with 1A2 serving as a secondary enzyme (Nielsen et al. 1996). Desmethylclomipramine is clomipramine's primary metabolite via demethylation by 2C19 and 3A4. The 2D6 enzyme performs subsequent hydroxylation. Fluvoxamine is a strong inhibitor of 1A2 and 2C19 and a moderate inhibitor of 3A4 (Christensen et al. 2002; von Moltke et al. 1995). Thus, when clomipramine was added in the presence of fluvoxamine, several P450 enzymes (3A4, 1A2, and 2C19) were impaired in their ability to efficiently metabolize clomipramine, resulting in blood levels of clomipramine+D-CMI that were significantly greater than they would have been had fluvoxamine not been present. When the fluvoxamine was replaced by venlafaxine, which is only a mild 2D6 inhibitor (Ball et al. 1997), the total level decreased more than threefold.

It is worth noting the vastly different ratios of clomipramine to D-CMI found in the above case. For the clomipramine plus fluvoxamine (50 mg/day) regimen, the ratio was roughly 7:1. For the clomipramine plus venlafaxine regimen, the ratio was roughly 1:4. This represents a roughly 28-fold ratio difference between these regimens! Since clomipramine is a potent serotonergic reuptake inhibitor and D-CMI is primarily a noradrenergic uptake inhibitor (Kelly and Myers 1990), these ratios may be important considerations as we refine our ability to reliably target pharmacologic interventions to specific illness profiles. This case suggests that at least some persons with OCD may respond more favorably to a "mixed action" antidepressant with both serotonergic and noradrenergic components rather than one focusing just on serotonin.

References

Ball SE, Ahern D, Scatina J, et al: Venlafaxine: in vitro inhibition of CYP2D6 dependent imipramine and desipramine metabolism; comparative studies with selected SSRIs, and effects on human hepatic CYP3A4, CYP2C9 and CYP1A2. Br J Clin Pharmacol 43(6):619–626, 1997

Christensen M, Tybring G, Mihara K, et al: Low daily 10-mg and 20-mg doses of fluvoxamine inhibit the metabolism of both caffeine (cytochrome P4501A2) and omeprazole (cytochrome P4502C19). Clin Pharmacol Ther 71:141–152, 2002

Kelly MW, Myers CW: Clomipramine: a tricyclic antidepressant effective in obsessive compulsive disorder. DICP 24(7–8):739–744, 1990

Nielsen KK, Flinois JP, Beaune P, et al: The biotransformation of clomipramine in vitro: identification of the cytochrome P450s responsible for the separate metabolic pathways. J Pharmacol Exp Ther 277(3):1659–1664, 1996

von Moltke LL, Greenblatt DJ, Court MH, et al: Inhibition of alprazolam and desipramine hydroxylation in vitro by paroxetine and fluvoxamine: comparison with other selective serotonin reuptake inhibitor antidepressants. J Clin Psychopharmacol 15(2):125–131, 1995

The Matrix

A 26-year-old man with schizophrenia was being stably maintained on haloperidol (Haldol), 10 mg/day, and benztropine (Cogentin), 2 mg/day. He then developed a seizure disorder, and his neurologist had him start taking phenytoin (Dilantin), 300 mg/day (blood level=16.7 μg/mL), which was effective in preventing further seizures. However, within 2 months of starting the phenytoin, the patient developed the delusion that he was Keanu Reeves and that everyone and everything around him was a computer-generated false reality whose sole purpose was to trick him into revealing the security and access codes for America's secret anthrax supply. To protect himself, the patient refused to leave his apartment, except for grocery shopping and visiting his psychiatrist. As the patient found this state of affairs quite upsetting, he agreed to be hospitalized. A trough haloperidol blood level drawn on admission was less than 2 ng/mL. Once his haloperidol dosage was increased to 40 mg/day, producing a blood level of 7.2 ng/mL, he began to experience a decrease in the intensity of his paranoid and grandiose delusions (S.C. Armstrong, personal communication, May 2002).

Discussion

This is an example of an inducer added to a substrate.

Haloperidol is metabolized by 3A4, 2D6, 1A2, and phase II glucuronidation (Desai et al. 2001; Kudo and Ishizaki 1999). Phenytoin is an inducer of 3A4, 2C9, and 2C19, as well as phase II glucuronidation (Chetty et al. 1998; Hachad et al. 2002; Spina and Perucca 2002). Thus, it appears that 3A4 is the P450 enzyme through which this interaction occurred. When the phenytoin was added, the amount of 3A4 that was available to metabolize the haloperidol increased, and this led to a lower haloperidol blood level, even though the dosage had not been changed. This decrease in the haloperidol blood level allowed for a reemergence of active delusions and the patient's accompanying clinical decompensation.

Part of this decrease in the haloperidol blood level may also be attributable to the induction of phase II glucuronidation. Although the specifics have not yet been well characterized, haloperidol also relies on phase II glucuronidation for a significant portion of its metabolism, and phenytoin is an inducer of the glucuronidation 1A4 enzyme (Hachad et al. 2002). Thus, the increased production and metabolic efficiency of this 1A4 enzyme may also have contributed to this inferred decrease in the haloperidol blood level.

Although there was no baseline haloperidol level for comparison, it would likely have been greater than 4 ng/mL at baseline, since that is considered the minimum effective blood level (although this threshold is far from reliable). The blood level was probably somewhat higher than 4 ng/mL, or else benztropine would probably not have been a necessary adjunct in treating the parkinsonian symptoms that occurred without benztropine while the patient was taking 10 mg/day of haloperidol. The increase in the haloperidol dosage to 40 mg/day appeared to compensate for phenytoin's induction of haloperidol's metabolism.

References

Chetty M, Miller R, Seymour MA: Phenytoin autoinduction. Ther Drug Monit 20(1):60–62, 1998

Desai HD, Seabolt J, Jann MW: Smoking in patients receiving psychotropic medications: a pharmacokinetic perspective. CNS Drugs 15(6):469–494, 2001

Hachad H, Ragueneau-Majlessi I, Levy RH: New antiepileptic drugs: review on drug interactions. Ther Drug Monit 24(1):91–103, 2002

Kudo S, Ishizaki T: Pharmacokinetics of haloperidol: an update. Clin Pharmacokinet 37(6):435–456, 1999

Spina E, Perucca E: Clinical significance of pharmacokinetic interactions between antiepileptic and psychotropic drugs. Epilepsia 43 (suppl 2):37–44, 2002

Vertigo

A 28-year-old man had been well maintained on phenobarbital (Luminal), 180 mg/day, for a long-standing seizure disorder. The patient then experienced a manic episode, leading to a psychiatric admission. The inpatient psychiatrist obtained a thorough history, revealing a lifelong pattern of abrupt mood swings, persistent social and occupational difficulties, numerous instances of episodic violence, and a family history of bipolar illness (father). The psychiatrist diagnosed the patient with rapid-cycling bipolar I disorder. In view of his symptom profile, and the fact that the patient's father had enjoyed a positive response to carbamazepine (Tegretol), he started the patient on carbamazepine, 200 mg bid, and the dosage was titrated to 800 mg/day over the next 8 days. Although the patient was calmer, he also was seen staggering while walking through the unit. He reported feeling sedated and confused, as well as having a persistent sensation that "the room is spinning around." The psychiatrist obtained a stat carbamazepine level, which was only 4.9 μg/mL. The phenobarbital level remained in the normal therapeutic range. Confused, he consulted one of his neurologist colleagues, who advised him either to discontinue the carbamazepine and consider trying lithium or even an atypical antipsychotic agent, or to retain the carbamazepine but discontinue the phenobarbital.

Discussion

This is an example of a substrate added to an inducer.

Carbamazepine is primarily a 3A4 substrate, although 1A2 and 2C9 play minor roles in carbamazepine's metabolism as well (Spina et al. 1996). Phenobarbital is an inducer of both 3A4 (Rautlin de la Roy et al. 1971) and 2C9 (El Adlouni et al. 2000). When a standard dosage of carbamazepine was administered in the presence of a 3A4 and 2C9 inducer (phenobarbital), the carbamazepine blood level was lower than expected, since the inducer led to greater production of 3A4 and 2C9 and a higher rate of metabolism of carbamazepine than would otherwise be expected (DeVane and Nemeroff 2002). This explains the carbamazepine blood level of 4.9 μg/mL, but not the vertigo, sedation, and confusion. These side effects were created by phenobarbital's induction of carbamazepine down a specific metabolic pathway (in this case, 3A4) that led to an excessive production of the neurotoxic carbamazepine-10,11-epoxide metabolite (Ayd 1995; Liu and Delgado 1995). In a sense, the presence of an inducer artificially "pulled" the substrate (carbamazepine) down a specific metabolic pathway to a greater degree than normal. Unless there is an effective physiologic compensation, this effect may lead to excessive production and accumulation of induced metabolites. In this case, the rate of production of carbamazepine-10,11-epoxide overwhelmed the body's ability to eliminate this metabolite via epoxide hydrolase, leading to just such accumulation and the side effects mentioned above. The consulting neurologist rightly suggested that this situation could be corrected by removing either the inducer (phenobarbital), which drove this process, or the substrate (carbamazepine), which supplied the precursor for this toxic metabolite.

Incidentally, 3A4 induction is not the only means of producing this toxic metabolite. For example, the addition of valproate to carbamazepine will also lead to the increased accumulation of carbamazepine-10,11-epoxide. However, this does not occur because of any valproate-related induction effects. Rather, valproate inhibits epoxide hydrolase, the enzyme that metabolizes carbamazepine-10,11-epoxide (Spina et al. 1996).

References

Ayd F: Lexicon of Psychiatry, Neurology, and the Neurosciences. Baltimore, MD, Williams & Wilkins, 1995, p 117

DeVane CL, Nemeroff CB: 2002 guide to psychotropic drug interactions. Primary Psychiatry 9(3):28–57, 2002

El Adlouni C, Pinelli E, Azemar B, et al: Phenobarbital increases DNA adduct and metabolites formed by ochratoxin A: role of CYP2C9 and microsomal glutathione-S-transferase. Environ Mol Mutagen 35(2):123–131, 2000

Liu H, Delgado MR: Interactions of phenobarbital and phenytoin with carbamazepine and its metabolites' concentrations, concentration ratios, and level/dose ratios in epileptic children. Epilepsia 36(3):249–254, 1995

Rautlin de la Roy Y DE, Beauchant G, Breuil K, et al: Reduction in the blood level of rifampicin by phenobarbital (in French). Presse Med 79(8):350, 1971

Spina E, Pisani F, Perucca E: Clinically significant pharmacokinetic drug interactions with carbamazepine: an update. Clin Pharmacokinet 31(3):198–214, 1996

Complications

A 54-year-old woman was hospitalized with a severe right femur fracture that occurred during a skiing accident. Her medical comorbidities included type II diabetes mellitus, controlled by diet alone, and a long-standing seizure disorder that had been well controlled with carbamazepine, 900 mg/day (blood level=9.2 µg/mL). In the weeks after surgery, her wound was not healing well and she began to experience intermittent fevers. A wound culture revealed infection with methicillin-resistant *Staphylococcus aureus* (MRSA). She was treated with 8 weeks of intravenous vancomycin, during which she was transitioned to home care with intravenous antibiotic treatment. The vancomycin successfully eradicated the MRSA, but the patient then developed a diffuse candidal rash. She was prescribed fluconazole (Diflucan), 200 mg on day 1 and 100 mg/day thereafter for at least 2 weeks. However, after 1 week she experienced increasing drowsiness, confusion, and incoordination. Three days later, she was stuporous and required readmission to the hospital. In the ER, her carbamazepine level was 19.4 µg/mL.

Discussion

This is an example of an inhibitor added to a substrate.

Carbamazepine is primarily a 3A4 substrate, although 1A2 and 2C9 make minor contributions to the metabolism of carbamazepine (Spina et al.

1996). Fluconazole is a strong 2C9 inhibitor (Cadle et al. 1994) and a moderate 3A4 inhibitor (Eap et al. 2002). With the addition of fluconazole, 3A4 and 2C9 were impaired in their ability to efficiently contribute to the metabolism of the carbamazepine. Since the activity of 1A2 was not sufficient to compensate for this effect, the 3A4 and 2C9 inhibition resulted in an increased blood level of carbamazepine, even though the carbamazepine dosage had remained constant (Nair and Morris 1999).

References

Cadle RM, Zenon GJ 3rd, Rodriguez-Barradas MC, et al: Fluconazole-induced symptomatic phenytoin toxicity. Ann Pharmacother 28(2):191–195, 1994

Eap CB, Buclin T, Baumann P: Interindividual variability of the clinical pharmacokinetics of methadone: implications for the treatment of opioid dependence. Clin Pharmacokinet 41(14):1153–1193, 2002

Nair DR, Morris HH: Potential fluconazole-induced carbamazepine toxicity. Ann Pharmacother 33(7–8):790–792, 1999

Spina E, Pisani F, Perucca E: Clinically significant pharmacokinetic drug interactions with carbamazepine: an update. Clin Pharmacokinet 31(3):198–214, 1996

Double Induction

A 27-year-old man with bipolar disorder had responded well to carbamazepine (Tegretol), 1,000 mg/day (blood level=9.6 μg/mL). He recently toured through Africa, and 6 months later his persistent cough and intermittent fevers led him to visit his internist, who eventually diagnosed active tuberculosis. He was started on rifampin (in the form of Rifadin), 600 mg/day. Three months later, he experienced a manic episode and was promptly hospitalized. A carbamazepine blood level was only 3.1 μg/mL, although the patient vehemently claimed that he had been compliant with his carbamazepine and rifampin. After consulting with the hospital pharmacist, the psychiatrist decided to add lithium (Eskalith) to his regimen, with the plan of eventually discontinuing the carbamazepine. In the meantime, the patient's rifampin blood level was also checked and was found to be at the extreme low end of the therapeutic range (4.8 μg/mL; therapeutic range: 4–32 μg/mL).

Discussion

This is a combined example of an inducer (rifampin) added to a substrate (carbamazepine) and a substrate (rifampin) added to an inducer (carbamazepine).

First, carbamazepine is primarily a 3A4 substrate, although 1A2 and 2C9 make minor contributions to its metabolism (Spina et al. 1996). Rifampin is an inducer of 3A4, 1A2, and 2C9 (Heimark et al. 1987; Strayhorn et al. 1997; Wietholtz et al. 1995; Zilly et al. 1977). When the rifampin was added to the carbamazepine, more 3A4, 1A2, and 2C9 were produced and thus available to more efficiently metabolize the carbamazepine (Pea and Furlanut 2001). This led to a decrease in the carbamazepine blood level to a subtherapeutic value, with the result that the patient's bipolar prophylaxis was impaired.

Second, rifampin is a 3A4 substrate (Strolin Benedetti and Dostert 1994), and carbamazepine is a 3A4 inducer (Spina et al. 1996). When the rifampin was prescribed at a standard dosage, it generated a lower than expected blood level because the carbamazepine had chronically increased the amount of 3A4 to greater than normal levels, thus increasing the ability to efficiently metabolize the rifampin. If the psychiatrist had wished to continue having the patient take both rifampin and carbamazepine, it would have been necessary to significantly increase the dosage of each agent. However, in order to avoid potential complications that could arise from such reciprocal induction, not to mention the difficulty of administering any other 3A4 substrate at a high enough dosage to be therapeutic, the psychiatrist instead opted to change the carbamazepine to lithium, with the plan to further add divalproex sodium (Depakote) if necessary.

References

Heimark LD, Gibaldi M, Trager WF, et al: The mechanism of the warfarin-rifampin drug interaction. Clin Pharmacol Ther 42(4):388–394, 1987

Pea F, Furlanut M: Pharmacokinetic aspects of treating infections in the intensive care unit: focus on drug interactions. Clin Pharmacokinet 40(11):833–868, 2001

Spina E, Pisani F, Perucca E: Clinically significant pharmacokinetic drug interactions with carbamazepine: an update. Clin Pharmacokinet 31(3):198–214, 1996

Strayhorn VA, Baciewicz AM, Self TH: Update on rifampin drug interactions, III. Arch Intern Med 157(21):2453–2458, 1997

Strolin Benedetti M, Dostert P: Induction and autoinduction properties of rifamycin derivatives: a review of animal and human studies. Environ Health Perspect 102 (suppl 9):101–105, 1994

Wietholtz H, Zysset T, Marschall HU, et al: The influence of rifampin treatment on caffeine clearance in healthy man. J Hepatol 22(1):78–81, 1995

Zilly W, Breimer DD, Richter E: Stimulation of drug metabolism by rifampicin in patients with cirrhosis or cholestasis measured by increased hexobarbital and tolbutamide clearance. Eur J Clin Pharmacol 11(4):287–293, 1977

Vigilance Always

A 35-year-old woman was being treated in an intensive outpatient setting for a significant worsening of her depressive symptoms, especially insomnia. Her medications included nefazodone (Serzone), 600 mg/day; lithium, 900 mg/day; amitriptyline (Elavil), 40 mg qhs; and sertraline (Zoloft), which had been increased from 100 mg/day to 150 mg/day in the previous few days. Both because of her insomnia and as a prudent measure, the psychiatrist checked her amitriptyline+nortriptyline blood level. To his surprise, the amitriptyline level was 472 ng/mL and the nortriptyline level was less than 20 ng/mL. He promptly discontinued the amitriptyline and started the patient on mirtazapine (Remeron) for insomnia and antidepressant augmentation, with lorazepam (Ativan) at night as needed for sleep. The nefazodone had been added to the amitriptyline roughly 6 months earlier (S. Khushalani, personal communication, July 2002).

Discussion

This is an example of an inhibitor added to a substrate.

Amitriptyline is a tertiary-amine TCA whose metabolism depends most on the intact functioning of 2C19, 3A4, and 2D6, with 1A2 serving as a secondary enzyme (Venkatakrishnan et al. 1998). Nortriptyline is amitriptyline's primary metabolite via demethylation by 2C19 and 3A4. The 2D6 enzyme performs subsequent hydroxylation. Nefazodone is a strong 3A4 inhibitor (von Moltke et al. 1996), and sertraline, at 150 mg/day, is a moderate 2D6 inhibitor (Alderman et al. 1997; Solai et al. 1997). These antidepressants (nefazodone and sertraline) markedly impaired the ability of 3A4 and 2D6 to metabolize the amitriptyline, thus leading to what was almost certainly an increase in the amitriptyline level (to 472 ng/mL), even though the dosage had not been changed. Certainly, this is a very high blood level for a comparatively low dosage of amitriptyline. Fortunately, the patient remained asymptomatic, and her amitriptyline was discontinued before she suffered any adverse consequences of her frankly toxic blood level.

The extremely low nortriptyline blood level indicates that strong 3A4 inhibition prevented much of the conversion (demethylation) of amitriptyline to nortriptyline by 3A4. However, 2D6 inhibition likely added to the direct elevation of the amitriptyline level by cutting off one of amitriptyline's accessory metabolic pathways (hydroxylation).

References

Alderman J, Preskorn SH, Greenblatt DJ, et al: Desipramine pharmacokinetics when coadministered with paroxetine or sertraline in extensive metabolizers. J Clin Psychopharmacol 17(4):284–291, 1997

Solai LK, Mulsant BH, Pollock BG, et al: Effect of sertraline on plasma nortriptyline levels in depressed elderly. J Clin Psychiatry 58(10):440–443, 1997

Venkatakrishnan K, Greenblatt DJ, von Moltke LL, et al: Five distinct human cytochromes mediate amitriptyline N-demethylation in vitro: dominance of CYP 2C19 and 3A4. J Clin Pharmacol 38(2):112–121, 1998

von Moltke LL, Greenblatt DJ, Harmatz JS, et al: Triazolam biotransformation by human liver microsomes in vitro: effects of metabolic inhibitors and clinical confirmation of a predicted interaction with ketoconazole. J Pharmacol Exp Ther 276(2):370–379, 1996

Do You Hear What I Hear?

A 32-year-old man with chronic paranoid schizophrenia had done reasonably well on haloperidol (Haldol), 10 mg/day, with benztropine (Cogentin), 2 mg/day. He had recently started a volunteer job, and the accompanying stress led to some new onset insomnia. He lived with his sister and nephew. The nephew had a seizure disorder and had been treated with phenobarbital, but this had been recently switched to valproate. However, there was still nearly a 1-month supply of unused phenobarbital in the house. The patient decided to take two tablets each night (each tab=30 mg) to help him get to sleep. The phenobarbital did ameliorate his onset insomnia, but he then experienced a gradual increase in his previous auditory hallucinations over the next several weeks. The increasingly frequent and intense barrage of voices—telling him, "Be careful! We're watching you!"—led to increasing paranoia, and he reported his difficulties to his psychiatrist. After taking a careful interval history, the psychiatrist 1) advised the patient to immediately discontinue the phenobarbital, 2) supplied the patient with lorazepam (Ativan) prn for insomnia, and 3) instructed him to increase his haloperidol to 20 mg/day and to report when EPS begin to emerge, at which point he would taper the haloperidol back to 10 mg/day. The patient carefully followed his instructions, his psychotic symptoms remitted over the next 3 weeks, and a hospitalization was avoided, albeit at the price of some transient stiffness and cogwheel rigidity.

Discussion

This is an example of an inducer added to a substrate.

Haloperidol is metabolized by 3A4, 2D6, 1A2, and phase II glucuronidation (Desai et al. 2001; Kudo and Ishizaki 1999). Phenobarbital is an inducer of 3A4 (among other P450 enzymes) (Rautlin de la Roy et al. 1971) and phase II glucuronidation (Hachad et al. 2002). The addition of phenobarbital led to an increase in the amount of 3A4, and possibly pertinent glucuronidation enzymes, that was produced and therefore available to metabolize the haloperidol. Over the several weeks during which this increased production of 3A4 (and possibly pertinent glucuronidation enzymes) occurred, there was a decrease in the blood level of haloperidol and a resulting recurrence of psychotic symptoms (Linnoila et al. 1980). The psychiatrist then recommended a temporary increase in the haloperidol dosage to compensate for these inductive effects. After the phenobarbital was discontinued, as the "extra" 3A4 (and possibly glucuronidation enzymes) that had been induced by phenobarbital "died off," there was an increase in the haloperidol blood level to a level greater than the patient's baseline, resulting in the emergence of EPS. The EPS served as a sign that the haloperidol could then be tapered back to its baseline dose. The lorazepam was supplied both to address the insomnia complaint and to prevent any possible sedative-hypnotic withdrawal state that might have occurred with the discontinuation of the phenobarbital.

References

Desai HD, Seabolt J, Jann MW: Smoking in patients receiving psychotropic medications: a pharmacokinetic perspective. CNS Drugs 15(6):469–494, 2001

Hachad H, Ragueneau-Majlessi I, Levy RH: New antiepileptic drugs: review on drug interactions. Ther Drug Monit 24(1):91–103, 2002

Kudo S, Ishizaki T: Pharmacokinetics of haloperidol: an update. Clin Pharmacokinet 37(6):435–456, 1999

Linnoila M, Viukari M, Vaisanen K, et al: Effect of anticonvulsants on plasma haloperidol and thioridazine levels. Am J Psychiatry 137(7):819–821, 1980

Rautlin de la Roy Y DE, Beauchant G, Breuil K, et al: Reduction of the blood level of rifampicin by phenobarbital (in French). Presse Med 79(8):350, 1971

Anticipatory Anxiety

A 45-year-old woman with panic disorder had been free from panic attacks for more than a year while taking sertraline (Zoloft), 100 mg/day. She had not responded to 50 mg/day during the initial determination of an effective dose. Three months ago, she was in a car accident and suffered a closed head injury. During her recovery process she had a few seizures, so the treating neurologist placed her on phenytoin (Dilantin), 300 mg/day (blood level=14.0 μg/mL), which did control her seizures. She continued to take her sertraline, but 2 weeks after starting the phenytoin, the patient began to experience a recurrence of her panic attacks. Her psychiatrist further increased the dosage of sertraline to 150 mg/day, but her panic attacks continued to increase in frequency and intensity. Her growing fear and anticipation of future panic attacks were further undermining efforts to treat her rapidly worsening panic disorder. In an effort to definitively interrupt this destructive spiral, the psychiatrist added clonazepam (Klonopin), 0.5 mg tid, which seemed to help in combination with the 150 mg/day of sertraline. The psychiatrist sought input from the hospital pharmacist. Following this consultation, the psychiatrist further increased the patient's sertraline to 200 mg/day, at which point he was able to taper and discontinue the clonazepam, and the patient remained free of panic attacks.

Discussion

This is an example of an inducer added to a substrate.

Sertraline, an inhibitor of serotonin reuptake, is itself metabolized at 2C9, 2C19, 2D6, and 3A4 (Greenblatt et al. 1999; Wang et al. 2001). Phenytoin is an inducer of 3A4, 2C9, and 2C19 (Chetty et al. 1998; Spina and Perucca 2002). Thus, the addition of the phenytoin led to an increase in the amounts of 3A4, 2C9, and 2C19 that were available to metabolize the sertraline, resulting in more efficient metabolism of the sertraline and a corresponding decrease in the blood level of sertraline. Studies have demonstrated that the addition of phenytoin to sertraline can significantly reduce sertraline levels (Pihlsgard and Eliasson 2002). In this case, the addition of phenytoin necessitated a doubling of the dosage of sertraline in order to produce a blood level of sertraline that was high enough to recapture a therapeutic response.

References

Chetty M, Miller R, Seymour MA: Phenytoin autoinduction. Ther Drug Monit 20(1): 60–62, 1998

Greenblatt DJ, von Moltke LL, Harmatz JS, et al: Human cytochromes and some newer antidepressants: kinetics, metabolism, and drug interactions. J Clin Psychopharmacol 19 (5, suppl 1):23S–35S, 1999

Pihlsgard M, Eliasson E: Significant reduction of sertraline plasma levels by carbamazepine and phenytoin. Eur J Clin Pharmacol 57(12):915–916, 2002

Spina E, Perucca E: Clinical significance of pharmacokinetic interactions between antiepileptic and psychotropic drugs. Epilepsia 43 (suppl 2):37–44, 2002

Wang JH, Liu ZQ, Wang W, et al: Pharmacokinetics of sertraline in relation to genetic polymorphism of CYP2C19. Clin Pharmacol Ther 70(1):42–47, 2001

Galactic Inconvenience

A 15-year-old with a psychotic depression first received risperidone (Risperdal), 0.25 mg/day, for 1 week with no ill effects, after which fluoxetine (Prozac), 20 mg/day, was added. Over the next 3 weeks, the patient gradually improved in terms of her psychotic and depressive symptoms. However, 1 month after the fluoxetine was started, she began to display galactorrhea. This problem persisted even when the dosage of risperidone was decreased to 0.25 mg every other day. The psychiatrist opted to substitute citalopram (Celexa), 20 mg/day, for the fluoxetine, and after 3 weeks her galactorrhea remitted. Curiously, at no time did she display any movement abnormalities (S. Ruths, personal communication, August 2002).

Discussion

This is an example of an inhibitor added to a substrate.

Risperidone is a substrate of 2D6 and 3A4 (DeVane and Nemeroff 2001). Fluoxetine (in concert with its active metabolite, norfluoxetine) is a strong 2D6 inhibitor and a moderate 3A4 inhibitor (Greenblatt et al. 1999; Stevens and Wrighton 1993). Thus, the addition of the fluoxetine significantly impaired the ability of 3A4 and 2D6 to efficiently metabolize the risperidone. This led to an increase in the blood level of risperidone, even though the dose of risperidone had not been increased. Since the increased blood level of risperidone resulted in more blockade of D_2 receptors, and since dopamine functions as "prolactin inhibitory hormone" in the tubero-infundibular dopamine pathway, there was an increase in serum prolactin, leading to galactorrhea. The absence of EPS was peculiar but likely indicates that her tubero-infundibular dopamine pathway was more sensitive to increased D_2 blockade than was her nigrostriatal dopamine pathway. When fluoxetine was discontinued and citalopram, which lacks any significant P450 inhibitory profile (Brosen and Naranjo 2001), was substituted, the reversal of 3A4 and 2D6 inhibition led to a return of the risperidone blood level to baseline. This led to a remission of her galactorrhea.

Studies in which fluoxetine was added to risperidone have revealed, on average, about a 75% increase in the blood level of the risperidone active moiety (Spina et al. 2002).

References

Brosen K, Naranjo CA: Review of pharmacokinetic and pharmacodynamic interaction studies with citalopram. Eur Neuropsychopharmacol 11(4):275–283, 2001

DeVane CL, Nemeroff CB: An evaluation of risperidone drug interactions. J Clin Psychopharmacol 21(4):408–416, 2001

Greenblatt DJ, von Moltke LL, Harmatz JS, et al: Human cytochromes and some newer antidepressants: kinetics, metabolism, and drug interactions. J Clin Psychopharmacol 19 (5, suppl 1):23S–35S, 1999

Spina E, Avenoso A, Scordo MG, et al: Inhibition of risperidone metabolism by fluoxetine in patients with schizophrenia: a clinically relevant pharmacokinetic drug interaction. J Clin Psychopharmacol 22(4):419–423, 2002

Stevens JC, Wrighton SA: Interaction of the enantiomers of fluoxetine and norfluoxetine with human liver cytochromes P450. J Pharmacol Exp Ther 266(2): 964–971, 1993

Sedated Akathisia

A 25-year-old man with schizoaffective disorder and a history of polysubstance abuse was being maintained on risperidone (Risperdal), 4 mg/day, and buspirone (BuSpar), 45 mg/day. He had achieved significant relief from his paranoid delusions and generalized anxiety on this regimen, but he had also begun to report increasing dysphoria, anhedonia, and decreased appetite. He had received the regimen of quetiapine (Seroquel) and fluoxetine (Prozac) in the past, and this had been effective, so the patient's psychiatrist decided to again add fluoxetine, 20 mg/day, to his regimen. Within 10 days, the patient was experiencing a peculiar state of alternating and mixed sedation and severe restlessness. He was frequently pacing the halls. At various times he would stop pacing, sit, abruptly fall asleep, then soon awaken and resume his pacing. Although the psychiatrist did not understand what had occurred, he discontinued the fluoxetine, and within 2 weeks these symptoms cleared (J. deLeon, personal communication, May 1999).

Discussion

This is an example of an inhibitor added to two substrates.

Risperidone is a substrate of 2D6 and 3A4 (DeVane and Nemeroff 2001), and buspirone is a 3A4 substrate (Kivisto et al. 1997). Fluoxetine (in concert with its active metabolite, norfluoxetine) is a strong 2D6 inhibitor and a moderate 3A4 inhibitor (Greenblatt et al. 1999; Stevens and Wrighton 1993). Thus, the addition of the fluoxetine significantly impaired the ability of 3A4 to metabolize the buspirone and the ability of both 3A4 and 2D6 to metabolize the risperidone, which led to increased blood levels of these substrates, even though their dosages had not been changed. The increased buspirone blood level yielded sedation, while the increased risperidone blood level caused akathisia. The discontinuation of the fluoxetine led to a reversal of the 3A4 and 2D6 inhibition, a return to the baseline blood levels of both risperidone and buspirone, and a remission of the sedation and akathisia.

Studies in which fluoxetine was added to risperidone have revealed, on average, about a 75% increase in the blood level of the risperidone active moiety (Spina et al. 2002).

References

DeVane CL, Nemeroff CB: An evaluation of risperidone drug interactions. J Clin Psychopharmacol 21(4):408–416, 2001

Greenblatt DJ, von Moltke LL, Harmatz JS, et al: Human cytochromes and some newer antidepressants: kinetics, metabolism, and drug interactions. J Clin Psychopharmacol 19 (5, suppl 1):23S–35S, 1999

Kivisto KT, Lamberg TS, Kantola T, et al: Plasma buspirone concentrations are greatly increased by erythromycin and itraconazole. Clin Pharmacol Ther 62(3):348–354, 1997

Spina E, Avenoso A, Scordo MG, et al: Inhibition of risperidone metabolism by fluoxetine in patients with schizophrenia: a clinically relevant pharmacokinetic drug interaction. J Clin Psychopharmacol 22(4):419–423, 2002

Stevens JC, Wrighton SA: Interaction of the enantiomers of fluoxetine and norfluoxetine with human liver cytochromes P450. J Pharmacol Exp Ther 266 (2):964–971, 1993

Chapter 7

Non-P450 Case Vignettes

These are cases in which the principal mode of drug interaction involves the P-glycoprotein transporter, phase II glucuronidation, or some other non-P450 pharmacokinetic issue. There are also cases in which the fundamental interaction is pharmacodynamic (interacting receptor-mediated end organ effects) rather than pharmacokinetic (interactions that affect absorption, metabolism, distribution, or excretion).

Secret Ingredient

A 55-year-old man with a history of alcohol dependence, in full remission, had been taking disulfiram (Antabuse), 250 mg/day, for the past 5 years. One morning, he awakened with crushing substernal chest pressure radiating to the left arm and jaw, as well as significant shortness of breath. He called 911 and was quickly transported to the nearest ER. Once he arrived, the staff found he was having a myocardial infarction complicated by congestive heart failure. Among other interventions, he was placed on a continuous nitroglycerin intravenous infusion. Within 15 minutes of the initiation of this infusion, he developed nausea, increased chest pain and dyspnea, tachycardia, diaphoresis, flushing, and a throbbing headache. Once the patient revealed that he was taking disulfiram, the nitroglycerin infusion was immediately discontinued and he was rushed to the medical ICU for more intensive monitoring and supportive care (K. Walters, personal communication, August 2002).

Discussion

This is an example of a substrate added to an inhibitor.

Nitroglycerin solutions for intravenous infusion contain a significant amount of ethyl alcohol (Nitroglycerin MICROMEDEX 1999). Disulfiram's main and intended function is to serve as a deterrent to the consumption of alcohol by virtue of its ability to inhibit aldehyde dehydrogenase, a key enzyme involved in the metabolism of ethyl alcohol. In the presence of alcohol, disulfiram's inhibition of this enzyme leads to an accumulation of acetaldehyde, resulting in the "disulfiram-alcohol reaction" (Kaplan and Sadock 1998) that this patient experienced. This interaction, even in generally healthy persons, has produced severe morbidity and even death, which has led to a decline in the popularity of disulfiram as a means of fostering abstinence from alcohol. When the ER physician recognized that this patient was having just such a reaction, he quickly discontinued the supply of the substrate (alcohol in the nitroglycerin infusion) and intensified the level of monitoring and care available to the patient.

References

Kaplan H, Sadock B: Kaplan and Sadock's Synopsis of Psychiatry: Behavioral Sciences/Clinical Psychiatry, 8th Edition. Baltimore, MD, Williams & Wilkins, 1998, p 1018

Nitroglycerin, in MICROMEDEX® Healthcare Series. Greenwood Village, CO, Micromedex Inc, 1974–1999

Contraceptive Convulsion

A 25-year-old woman with a long-standing seizure disorder had been seizure-free for 5 years while taking lamotrigine (Lamictal), 200 mg/day. She then began a serious relationship with a boyfriend, and she visited her gynecologist in order to start taking oral contraceptives. After the appropriate examinations and tests, she started taking a standard oral contraceptive containing ethinylestradiol. Events proceeded uneventfully until she had a grand mal seizure 3 weeks later (J.R. Oesterheld, personal communication, May 2002).

Discussion

This is an example of an inducer added to a substrate.

Lamotrigine is primarily metabolized through phase II glucuronidation, specifically by the uridine 5′-diphosphate glucuronosyltransferase (UGT) 1A4 enzyme (Hiller et al. 1999). Ethinylestradiol is an inducer of UGT1A4

(Sabers et al. 2001). With the introduction of the ethinylestradiol, more UGT1A4 was produced and thus available to more efficiently metabolize the lamotrigine. This led to a decrease in the blood level of lamotrigine, with the result that the patient experienced her first seizure in more than 5 years. One study demonstrated that adding oral contraceptives in patients taking lamotrigine led to a mean 50% decrease in lamotrigine blood levels (Sabers et al. 2001).

Phase II glucuronidation seems to function in a manner analogous to P450 oxidative metabolism in that there are glucuronidation enzymes, each of which has its own substrates, inhibitors, and inducers. The analogy is not precise in several ways. There may be more "crossover" metabolism among closely related glucuronidation enzymes than with P450 enzymes, and the specific nature of the interactions is clearly different, the kinetics are different, and so forth. However, there is an emerging appreciation for the importance and clinical relevance of the phase II component of drug metabolism.

References

Hiller A, Nguyen N, Strassburg CP, et al: Retigabine N-glucuronidation and its potential role in enterohepatic circulation. Drug Metab Dispos 27(5):605–612, 1999

Sabers A, Buchholt JM, Uldall P, et al: Lamotrigine plasma levels reduced by oral contraceptives. Epilepsy Res 47(1–2):151–154, 2001

Rash Decision (I)

The 25-year-old woman in the previous case (see "Contraceptive Convulsion") eventually had her lamotrigine (Lamictal) dosage increased from 200 mg/day to 400 mg/day to compensate for the induction of the lamotrigine by her oral contraceptives. Two more years passed uneventfully, with no recurrence of seizure activity, until she developed a deep venous thrombosis in her right thigh during a transatlantic plane flight. After the patient had received appropriate acute treatment, her gynecologist informed her that her oral contraceptives could have predisposed her to this event and that they should be discontinued at this point. The patient agreed, and she promptly stopped taking the contraceptives. Over the next several weeks, the patient felt more pervasively fatigued and forgetful, but she attributed this to more stress at work of late. One month after discontinuing the contraceptives, however, she developed a diffuse, red rash along her upper arms, shoulders, and upper back. She promptly informed her neurologist of this rash, as she had been previously advised, and he instructed her to discontinue the lamotrigine and report to his office immediately so that he could examine the rash and plan how to proceed (J.R. Oesterheld, personal communication, May 2002).

Discussion

This is an example of reversal of induction.

As mentioned in the previous case ("Contraceptive Convulsion"), lamotrigine is primarily metabolized through phase II glucuronidation, specifically by the uridine 5′-diphosphate glucuronosyltransferase (UGT) 1A4 enzyme (Hiller et al. 1999), and ethinylestradiol is an inducer of UGT1A4 (Sabers et al. 2001). With the discontinuation of the ethinylestradiol, the "extra" UGT1A4 that had caused the lamotrigine to be more efficiently metabolized gradually "died off," and UGT1A4 returned to its baseline level of functioning. This led to a precipitous increase in her blood level of lamotrigine, which caused her fatigue and the emergence of a lamotrigine rash. Lamotrigine rashes generally warrant immediate discontinuation of this medication, as they can be the precursor of a full Stevens-Johnson syndrome or toxic epidermal necrolysis.

References

Hiller A, Nguyen N, Strassburg CP, et al: Retigabine N-glucuronidation and its potential role in enterohepatic circulation. Drug Metab Dispos 27(5):605–612, 1999

Sabers A, Buchholt JM, Uldall P, et al: Lamotrigine plasma levels reduced by oral contraceptives. Epilepsy Res 47(1–2):151–154, 2001

Premature Crossover

A 23-year-old woman with a history of melancholic major depression was given fluoxetine (Prozac), which was titrated to a dosage of 60 mg/day, with little clinical benefit. Her psychiatrist decided to embark on a trial of venlafaxine (Effexor). Three days after abruptly discontinuing the fluoxetine, the patient was started on venlafaxine XR, 150 mg/day. Within another 3 days, the patient experienced disturbing myoclonus along with flushing, diarrhea, and fever. The venlafaxine was discontinued, and she was admitted to the medical floor of a nearby hospital, where she received supportive treatment with antipyretics and lorazepam (Ativan). She was discharged without complications 2 days later.

She did eventually try venlafaxine again, but she waited another 4 weeks before doing so (J.R. Oesterheld, personal communication, May 2002).

Discussion

This is a combined example of a substrate added to an inhibitor and excessive pharmacodynamic synergy.

Fluoxetine is a potent selective serotonin reuptake inhibitor (SSRI) with an average half-life of 1–3 days (Prozac [package insert] 2001). However, fluoxetine's major metabolite, norfluoxetine, is an equally potent serotonin reuptake inhibitor, but its average half-life is 17 days (Brunswick et al. 2001). Venlafaxine, at 150 mg/day, also functions as a potent serotonin reuptake inhibitor. Thus, the 5–6 days between the discontinuation of the fluoxetine and the time it took for venlafaxine to reach steady state was not sufficient to prevent the excessive additive effects of fluoxetine's and, especially, norfluoxetine's still considerable serotonin reuptake inhibitor activity combined with that of venlafaxine. The end result was a mild to moderate central serotonin syndrome leading to an avoidable hospitalization.

A likely contributor to this serotonin syndrome was the pharmacokinetic interaction between fluoxetine and venlafaxine. Venlafaxine is predominantly a substrate of 2D6 (Effexor [package insert] 2002), and fluoxetine/norfluoxetine is a potent 2D6 inhibitor (Stevens and Wrighton 1993). Thus, the considerable amount of norfluoxetine that was present when venlafaxine reached steady state was still able to strongly inhibit 2D6, leading to a greater than expected blood level of venlafaxine and thereby increasing the likelihood of an adverse event such as a serotonin syndrome.

References

Brunswick DJ, Amsterdam JD, Fawcett J, et al: Fluoxetine and norfluoxetine plasma levels after discontinuing fluoxetine therapy. J Clin Psychopharmacol 21(6):616–618, 2001

Effexor (package insert). Philadelphia, PA, Wyeth Laboratories, April 2002

Prozac (package insert). Eli Lilly/Dista Pharmaceuticals, February 2001

Stevens JC, Wrighton SA: Interaction of the enantiomers of fluoxetine and norfluoxetine with human liver cytochromes P450. J Pharmacol Exp Ther 266(2): 964–971, 1993

A Little Goes a Long Way

A 36-year-old woman with bipolar I disorder had been stably maintained on lamotrigine (Lamictal), 200 mg/day (blood level=2.5 μg/mL), for the past 18 months, which qualified as a long period of stability for this patient. Some persistent dysphoria, anhedonia, and insomnia began to develop, so her psychiatrist decided to introduce sertraline (Zoloft) at a low dosage (25 mg/day) into her regimen. Over the next several days, the patient developed increasing lethargy and confusion. When she reported this to her psychiatrist, he ordered another lamotrigine blood level, which was found to be 5.1 μg/mL. After consulting with the affiliated hospital's pharmacist, the psychiatrist discontinued the sertraline and started the patient on citalopram (Celexa), with no further difficulties of this nature.

Discussion

This is an example of an inhibitor added to a substrate.

Lamotrigine is metabolized primarily through phase II glucuronidation, specifically by the uridine 5′-diphosphate glucuronosyltransferase (UGT) 1A4 enzyme (Hiller et al. 1999). Sertraline, seemingly unique among the SSRIs, is an inhibitor of UGT1A4 (Kaufman and Gerner 1998). Thus, the addition of sertraline (and at a low dosage at that) significantly impaired the ability of UGT1A4 to metabolize the lamotrigine. This led to a sharp increase in the blood level of lamotrigine, a doubling in this case (Kaufman and Gerner 1998). Since the final lamotrigine blood level was not terribly high (5.1 μg/mL), the patient's resulting symptoms were not too dramatic or distressing. However, it is worth noting that only 25 mg/day of sertraline was able to double the lamotrigine blood level. This would have been more expected if valproate had been the added agent, yet sertraline appears to be every bit as effective in elevating lamotrigine blood levels as valproate (see "Rash Decision [I]," p. 201, for further discussion of these interactions).

References

Hiller A, Nguyen N, Strassburg CP, et al: Retigabine N-glucuronidation and its potential role in enterohepatic circulation. Drug Metab Dispos 27(5):605–612, 1999

Kaufman KR, Gerner R: Lamotrigine toxicity secondary to sertraline. Seizure 7(2):163–165, 1998

Cholestatic Catastrophe

A 45-year-old woman whose atypical major depression was being well managed with phenelzine (Nardil), 60 mg/day, was admitted to the surgical floor of a hospital for cholecystitis with severe, colicky abdominal pain. Her surgical intern ordered meperidine (Demerol), 50 mg im every 4 hours prn for pain, since he had "learned" that meperidine was less likely than morphine to contract the sphincter of Oddi and thus exacerbate her cholestatic pain. Within 5 hours of receiving her first (and only) injection of meperidine, she became acutely confused and agitated and developed a fever of 105.9°. Despite all supportive attempts, the patient expired over the next 2 hours.

Discussion

This is an example of a "substrate" added to an inhibitor.

In this case, severe serotonin syndrome was caused by the interaction of phenelzine (a monoamine oxidase inhibitor, or MAOI) and meperidine, a narcotic analgesic that acts as a "serotomimetic agent" (Bodner et al. 1995). Since this property of meperidine led to a rapid increase in synaptically available serotonin, and since phenelzine inhibits the breakdown of all monoamines (including serotonin), the combination of these two agents led to an acute and severe accumulation of serotonin, culminating in an acute central serotonin syndrome (Sporer 1995). This case mirrors the famous Libby Zion case, in which Ms. Zion, who was regularly taking phenelzine, received meperidine in an ER for shaking chills. She developed a fever of 107.6° and died within 6 hours of her arrival in that hospital (Asch and Parker 1988).

References

Asch DA, Parker RM: The Libby Zion Case: one step forward or two steps backward? N Engl J Med 318(12):771–775, 1988

Bodner RA, Lynch T, Lewis L, et al: Serotonin syndrome. Neurology 45(2):219–223, 1995

Sporer KA: The serotonin syndrome: implicated drugs, pathophysiology and management. Drug Saf 13(2):94–104, 1995

Fibrillations

A 65-year-old widower with atrial fibrillation was taking digoxin (Lanoxin), 0.25 mg/day (blood level=0.8 ng/mL), for rate control and one aspirin each day as an anticlotting agent. This regimen had been generally effective, producing average resting heart rates in the range of 60 to 80 beats per minute. After his pet bull terrier died, the man became more dysphoric and demoralized. He did not want to discuss his feelings with any of his doctors, so he instead opted to visit the local drugstore and purchase St. John's wort *(Hypericum perforatum)*, which he then began to consume per the pharmacist's instructions. In about 2 weeks, he noticed the sensation that his heart was racing, and his pulse felt to him like it was faster than 100 beats per minute. He called his doctor, who instructed him to call 911 immediately. The heart monitor in the ambulance confirmed that he was again in rapid atrial fibrillation, and his digoxin blood level in the ER had declined to 0.5 ng/mL.

Discussion

This is an example of an inducer added to a substrate.

Digoxin is a substrate of the P-glycoprotein transport protein (Pauli-Magnus et al. 2001), and St. John's wort is an inducer of P-glycoprotein (Hennessy et al. 2002). Thus, the addition of the St. John's wort led to the increased production of this transporter, which was therefore more active in extruding digoxin from enterocytes and back into the gut lumen, where it was excreted rather than absorbed. This caused a corresponding decline in the digoxin blood level, rendering the digoxin less effective in providing rate control of the patient's atrial fibrillation. Studies have demonstrated a 33% decrease in digoxin blood levels when St. John's wort is added for at least 10 days (Johne et al. 1999).

References

Hennessy M, Kelleher D, Spiers JP, et al: St Johns wort increases expression of P-glycoprotein: implications for drug interactions. Br J Clin Pharmacol 53(1): 75–82, 2002

Johne A, Brockmoller J, Bauer S, et al: Pharmacokinetic interaction of digoxin with an herbal extract from St. John's wort *(Hypericum perforatum)*. Clin Pharmacol Ther 66(4):338–345, 1999

Pauli-Magnus C, Murdter T, Godel A, et al: P-glycoprotein–mediated transport of digitoxin, alpha-methyldigoxin and beta-acetyldigoxin. Naunyn Schmiedebergs Arch Pharmacol 363(3):337–343, 2001

St. Serotonin

A 28-year-old woman with dysthymic disorder had been prescribed paroxetine (Paxil), 20 mg/day, by her primary care physician. She had not yet enjoyed a positive response to this medication, so she unilaterally decided to augment the paroxetine with St. John's wort *(Hypericum perforatum).* Within 5 days, she experienced increasing nausea, agitation, ataxia, diaphoresis, muscle stiffness, and tremor. She reported these symptoms to her physician, who then advised her to call 911 and report to the nearest ER. Once there, she was admitted to the medical floor and treated with muscle relaxants, lorazepam (Ativan), and antipyretics for the fevers that she had begun to experience on admission. She was discharged in good condition 3 days later. The primary care physician obtained preauthorization from the patient's HMO for her to obtain subsequent psychiatric care from a psychiatrist.

Discussion

This is an example of a central serotonin syndrome, moderate in severity, arising from two complementary mechanisms.

First, both paroxetine (an SSRI) and St. John's wort have significant abilities to inhibit serotonin reuptake. When these agents are combined, these effects are additive, which contributed significantly to the development of the serotonin syndrome. Second, St. John's wort also acts as a weak monoamine oxidase inhibitor (MAOI) (Gnerre et al. 2001). Since serotonin is a monoamine, the St. John's wort therefore inhibited the metabolic breakdown of serotonin by monoamine oxidase. This further increased the level of serotonin activity, and this increase synergized with the additive serotonin reuptake inhibition of the two agents to produce a central serotonin syndrome (DeVane and Nemeroff 2002).

References

DeVane CL, Nemeroff CB: 2002 guide to psychotropic drug interactions. Primary Psychiatry 9(3):28–57, 2002

Gnerre C, von Poser GL, Ferraz A, et al: Monoamine oxidase inhibitory activity of some *Hypericum* species native to south Brazil. J Pharm Pharmacol 53(9): 1273–1279, 2001

The Tremulous Triathlete

A 31-year-old triathlete with bipolar I disorder, which was being well controlled with lithium (Eskalith), 900 mg/day (blood level=0.8 mEq/L), developed severe tendinitis in his right ankle 2 weeks before a major triathlon, for which he had been training for months. His orthopedist prescribed a nonsteroidal anti-inflammatory drug (NSAID), indomethacin (Indocin), 50 mg tid, until after the triathlon. Over the next several days, the patient noted that he felt increasingly "spacey." He also noted that he was urinating more frequently. He was well aware of the importance of good hydration while taking lithium, but between sweating during his training and his frequent urination, the important task of hydration became harder to maintain. When he developed a frank tremor that made it difficult for him to feed himself or write, he decided to contact his psychiatrist. She asked the patient about any recent changes in his medication regimen. On hearing that he was taking indomethacin on a regular basis, she had him report for an immediate lithium blood level, which was found to be 1.5 mEq/L.

Discussion

This is an example of mild NSAID-induced lithium toxicity.

Lithium is entirely renally excreted, and therefore a variety of medications that alter renal function can alter the rate of lithium excretion. All NSAIDs (including indomethacin), with the exception of aspirin and sulindac (Clinoril), will often elevate lithium blood levels (Ragheb 1990). The change in lithium blood level can be quite variable and difficult to predict, so close observation and frequent lithium levels are warranted. The proposed mechanism for this interaction is that NSAIDs inhibit prostaglandin synthesis in the kidney, thus interfering with the excretion of lithium (Imbs et al. 1997). In this case, the patient's mild dehydration also likely contributed to his emerging lithium toxicity. Whenever a patient who is taking lithium is placed on a standing regimen of a nonaspirin, nonsulindac NSAID, lithium levels should be closely monitored and, perhaps, lithium doses should even be slightly lowered in anticipation of this interaction, especially in those individuals whose maintenance lithium blood levels are on the high end of the maintenance range (0.8–1.0 mEq/L).

References

Imbs JL, Barthelmebs M, Danion JM, et al: Mechanisms of drug interactions with renal elimination of lithium (in French). Bull Acad Natl Med 181(4):685–697, 1997

Ragheb M: The clinical significance of lithium–nonsteroidal anti-inflammatory drug interactions. J Clin Psychopharmacol 10(5):350–354, 1990

Baffling Hypertension

A 47-year-old man with schizoaffective disorder, bipolar type, was being unsuccessfully treated with the following medications: fluphenazine (Prolixin), 10 mg/day; benztropine (Cogentin), 1 mg/day; lithium carbonate, 900 mg/day; and phenelzine, 15 mg tid. Despite this broad-spectrum regimen, he continued to endorse paranoid delusions that he was being followed and harassed by the FBI and CIA and that they had implanted a tracking device behind his left eye. He was becoming preoccupied with enucleating himself and thus removing this device, so the psychiatrist decided to add clozapine (Clozaril) to his regimen, titrating to a dosage of 400 mg/day. Since the patient was now receiving two agents that were notorious for producing orthostatic hypotension (phenelzine and clozapine), the psychiatrist was vigilant in monitoring the patient's orthostatic blood pressure readings twice each day. After 1 week at this dosage of clozapine, however, the patient's blood pressure readings began to climb. In a few days, his blood pressure had risen from a typical 115/80 mm Hg to 180/105 mm Hg. The clozapine was gradually tapered and discontinued, and the patient's blood pressure returned to baseline (K. Walters, personal communication, July 2002).

Discussion

This is an example of a mild MAOI-clozapine interaction–induced hypertensive episode.

By virtue of α_1-adrenergic blockade, both phenelzine and clozapine often produce orthostatic hypotension (Kaplan and Sadock 1998). However, phenelzine is a monoamine oxidase inhibitor (MAOI), which can interact with agents that increase the availability of catecholamines (epinephrine, norepinephrine, and dopamine) to produce hytertensive crises. Clozapine, in addition to its multiple dopamine and serotonin receptor blockade effects, also blocks central α_2-adrenergic receptors (Kaplan and Sadock 1998). Insofar as this is the noradrenergic system autoreceptor, blockade of this receptor by clozapine will lead to increased presynaptic release of norepinephrine. In one case series, Breier et al. (1994) found that the administration of clozapine led to a mean 471% increase in plasma norepinephrine levels. As an MAOI, phenelzine inhibited the breakdown of this increased amount of norepinephrine, leading to an increase in blood pressure that was likely, and fortunately, mitigated by the hypotension-producing α_1 blockade profiles of both compounds. The surprise for the treatment team lay in the fact that the combination of two typically orthostasis-producing agents resulted in a hypertensive state.

References

Breier A, Buchanan RW, Waltrip RW 2nd, et al: The effect of clozapine on plasma norepinephrine: relationship to clinical efficacy. Neuropsychopharmacology 10(1):1–7, 1994

Kaplan H, Sadock B: Kaplan and Sadock's Synopsis of Psychiatry: Behavioral Sciences/Clinical Psychiatry, 8th Edition. Baltimore, MD, Williams & Wilkins, 1998, pp. 945, 1072

The Curse of Zoster

A 45-year-old man began to experience severe and debilitating postherpetic neuralgia pain (right-sided ophthalmic and maxillary distribution) about 3 years ago. After several unsuccessful attempts at treatment, the neurologist eventually prescribed morphine as MS Contin, 100 mg bid, which controlled the patient's extreme pain. Two months ago, the patient reported persistent cough, malaise, and fevers to his internist, and he was eventually diagnosed with tuberculosis. Among other medications, he was placed on rifampin (in the form of Rifadin). One week after the rifampin was started, he began to experience the reemergence of his neuralgia pain, and MS Contin, 100 mg bid, was again prescribed to address this. One week later, the pain was almost as bad as it had been before the MS Contin was started. The neurologist immediately began to titrate the MS Contin dose upward. The patient reachieved his baseline of easily tolerated pain only after an MS Contin dose of 180 mg bid had been reached.

Discussion

This is an example of an inducer added to a substrate.

Morphine is a substrate of the P-glycoprotein transport protein (Wandel et al. 2002), and rifampin is an inducer of P-glycoprotein (Geick et al. 2001). Thus, the addition of rifampin led to an increased production of this transporter, which was therefore more active in extruding morphine from enterocytes and back into the gut lumen, where it was excreted rather than absorbed. This caused a corresponding decline in the blood level of morphine, rendering the original dose much less effective in controlling the patient's postherpetic neuralgia pain. One study demonstrated a 28% decrease in the morphine AUC (area under the curve), a 41% decrease in the maximum blood level of morphine, and a loss of analgesic efficacy after rifampin was added for a 2-week period (Fromm et al. 1997).

References

Fromm MF, Eckhardt K, Li S, et al: Loss of analgesic effect of morphine due to coadministration of rifampin. Pain 72(1–2):261–267, 1997

Geick A, Eichelbaum M, Burk O: Nuclear receptor response elements mediate induction of intestinal MDR1 by rifampin. J Biol Chem 276(18):14581–14587, 2001

Wandel C, Kim R, Wood M, et al: Interaction of morphine, fentanyl, sufentanil, alfentanil, and loperamide with the efflux drug transporter P-glycoprotein. Anesthesiology 96(4):913–920, 2002

DI Difficulties

A 23-year-old man was diagnosed with bipolar I disorder, following his first unequivocal manic episode. He was initially treated with lithium (Eskalith), 900 mg/day (blood level=0.9 mEq/L), in no small part because his father, who also had bipolar disorder, had enjoyed a profound and sustained response to this medication over the years. Although the lithium was as effective for the patient as it had been for his father, he was experiencing a side effect his father had not experienced. His urinary output was about 2.5 liters per day, requiring the patient to engage in an onerous amount of fluid repletion. He also experienced constant thirst, the inconvenience of hourly trips to the bathroom, interrupted sleep, and even occasional nocturnal enuresis. He asked his psychiatrist for help with this problem, and his psychiatrist added hydrochlorothiazide, 50 mg/day. Although this thiazide diuretic exerted its anticipated paradoxical effect of decreasing the patient's urine production, over the next 3 days he experienced increasing tremor, ataxia, dysarthria, and sedation. The psychiatrist instructed the patient to go to the ER to obtain a lithium blood level, which was 1.4 mEq/L.

Discussion

This is an example of lithium toxicity caused by a sodium-depleting agent.

Hydrochlorothiazide causes increased sodium excretion at the distal convoluted tubule of the nephron. Grossly speaking, the kidney tries to compensate for this effect by retaining sodium at other portions of the nephron. Since lithium and sodium are both monovalent cations, the kidney confuses the two ions and thus acutely decreases its clearance of lithium as well (Kaplan and Sadock 1998). This retention of lithium then produces an increase in the lithium blood level, usually by roughly 25%–40% (Finley et al. 1995). In this case, since the baseline lithium level was relatively high (0.9 mEq/L), the resulting blood level produced a state of mild lithium toxicity.

A better choice for decreasing urine output in lithium-induced diabetes insipidus (DI) would have been amiloride (Midamor), which accomplishes the same decrease in urine production but generally without significant changes in lithium or potassium levels (Batlle et al. 1985).

References

Batlle DC, von Riotte AB, Gaviria M, et al: Amelioration of polyuria by amiloride in patients receiving long-term lithium therapy. N Engl J Med 312(7):408–414, 1985

Finley PR, Warner MD, Peabody CA: Clinical relevance of drug interactions with lithium. Clin Pharmacokinet 29(3):172–191, 1995

Kaplan H, Sadock B: Kaplan and Sadock's Synopsis of Psychiatry: Behavioral Sciences/Clinical Psychiatry, 8th Edition. Baltimore, MD, Williams & Wilkins, 1998, p 1052

Nauseated Nanny

A 40-year-old nanny with recurrent major depression had done well for the past 9 months on the regimen of citalopram (Celexa), 20 mg/day; venlafaxine (Effexor), 300 mg/day (added 10 months ago); and lithium, 1,050 mg/day (blood level=0.85 mEq/L). During a routine appointment with her internist, she was found to have a blood pressure of 170/100 mm Hg. After discussing various options with the patient, the internist prescribed lisinopril (Zestril), which was quickly titrated to a dosage of 30 mg/day. Within 1 week of reaching this dosage, however, the patient contacted both her psychiatrist and her internist to report a coarse intention tremor, blurry vision, and persistent nausea with vomiting once each day. The psychiatrist suggested that she report to the nearest ER for a lithium level, which was found to be 1.4 mEq/L (Baldwin and Safferman 1990).

Discussion

This is an example of mild lithium toxicity caused by a drug that inhibits the production of aldosterone.

Lithium is excreted primarily by the kidney. Lisinopril is an angiotensin-converting enzyme (ACE) inhibitor. (ACE is the enzyme that converts angiotensin I into angiotensin II.) Angiotensin II, in addition to being a vasoconstrictor, promotes the release of the hormone aldosterone, which acts at the kidney to retain sodium and excrete potassium. Since ACE inhibitors ultimately inhibit the release of aldosterone, they shift the renal handling of sodium in the direction of decreased retention and increased excretion. Sodium and lithium compete for reabsorption at the proximal tubule of the nephron. With relatively less sodium being retained because of ACE inhibitor–induced decreases in aldosterone release, lithium is better able to compete for reabsorption, leading to increased lithium retention and possible toxicity (Finley et al. 1996).

References

Baldwin CM, Safferman AZ: A case of lisinopril-induced lithium toxicity. DICP 24(10):946–947, 1990

Finley PR, O'Brien JG, Coleman RW: Lithium and angiotensin-converting enzyme inhibitors: evaluation of a potential interaction. J Clin Psychopharmacol 16(1):68–71, 1996

Resistance

A 45-year-old man with panic disorder had been free of panic attacks for 3 years while taking paroxetine (Paxil), 40 mg/day. One summer, he was hospitalized after his right arm and hand were badly mutilated by a lawnmower he was attempting to repair. In the week following surgery, he began to experience intermittent fevers and felt especially weak. Blood cultures were positive for methicillin-resistant *Staphylococcus aureus* (MRSA). The patient was placed on a therapeutic dose of vancomycin, and his fevers abated in a few days. However, 1 week after the vancomycin was started, his fevers and a renewed feeling of malaise and nausea returned. Another set of blood cultures now revealed a vancomycin-resistant *Enterococcus faecium* bacteremia. The vancomycin was discontinued, and he was placed on linezolid (Zyvox), 600 mg iv every 12 hours. The patient was still taking his paroxetine at the regular dosage. Within 2 days, the patient was acutely delirious, febrile to 104°, hypertensive, vomiting, and experiencing myoclonus. The patient was immediately taken to the ICU for more intensive care and monitoring (Wigen and Goetz 2002).

Discussion

This is an example of an inhibitor added to a substrate.

Paroxetine is a potent selective serotonin reuptake inhibitor (SSRI). SSRIs increase serotonergic neurotransmission and overall availability of serotonin. Linezolid, in addition to being an antibiotic, is a reversible but nonselective monoamine oxidase inhibitor (MAOI), and as such it inhibits the breakdown of serotonin. The combination of these two mechanisms of action resulted in a toxic accumulation of serotonin, manifesting as an acute serotonin syndrome (Wigen and Goetz 2002) with autonomic instability and delirium.

References

Wigen CL, Goetz MB: Serotonin syndrome and linezolid. Clin Infect Dis 34(12): 1651–1652, 2002

Rash Decision (II)

A 30-year-old woman with a history of bipolar I disorder had been stably maintained on lamotrigine (Lamictal), 250 mg/day, and lithium, 1,200 mg/day (blood level=0.9 mEq/L). However, she was recently hospitalized because of a breakthrough manic episode. Because of her aversion to the concept of antipsychotic medication, she declined the offer of an adjunctive atypical antipsychotic and instead consented to the addition of divalproex sodium (Depakote), titrated to a dosage of 1,000 mg/day (blood level=92 μg/mL). Her manic symptoms gradually remitted, but within 1 week she reported feeling dizzy, sedated, and "slowed." Additionally, a diffuse rash appeared on her upper arms. The psychiatrist immediately discontinued the lamotrigine, and in a few days the rash began to fade and the patient continued to improve clinically (N.B. Sandson, self-report, February 2001).

Discussion

This is an example (primarily) of an inhibitor added to a substrate.

Lamotrigine is metabolized primarily through phase II glucuronidation, specifically by the 1A4 enzyme (Hiller et al. 1999). Divalproex is an inhibitor of this same glucuronidation enzyme (Hachad et al. 2002). Thus, the addition of the divalproex significantly impaired the ability of the 1A4 enzyme to efficiently metabolize the lamotrigine, which led to an increase in the blood level of lamotrigine, even though the dosage of lamotrigine had not been changed. This rapid rise in the lamotrigine blood level caused the patient to experience the symptoms of sedation and mild confusion, as well as the emergence of a rash. Lamotrigine rashes generally warrant immediate discontinuation of this medication, as they can be the precursor of a full Stevens-Johnson syndrome or toxic epidermal necrolysis.

In addition to the elevation of lamotrigine blood levels by divalproex, it appears that adding lamotrigine to divalproex will consistently decrease divalproex blood levels by roughly 25% (DeVane and Nemeroff 2002). The likely explanation is that lamotrigine seems to have some weak induction effects on glucuronidation enzymes (not as yet formally characterized), and the glucuronidative metabolism of divalproex (again, not yet well characterized) is accordingly enhanced, yielding a modest decrease in the blood levels of divalproex.

References

DeVane CL, Nemeroff CB: 2002 guide to psychotropic drug interactions. Primary Psychiatry 9(3):28–57, 2002

Hachad H, Ragueneau-Majlessi I, Levy RH: New antiepileptic drugs: review on drug interactions. Ther Drug Monit 24(1):91–103, 2002

Hiller A, Nguyen N, Strassburg CP, et al: Retigabine N-glucuronidation and its potential role in enterohepatic circulation. Drug Metab Dispos 27(5):605–612, 1999

More Than He Bargained For

A 33-year-old man with recently diagnosed bipolar I disorder had remained essentially euthymic for the past 4 months while taking lithium, 900 mg/day (blood level=0.75 mEq/L). However, he consistently complained that he felt "blunted" and "less alive" than in his pre-lithium days, when he would frequently cycle into hypomanic episodes. His psychiatrist tried to counsel him and help him accept and adapt to the changes that treatment was producing for him, including the "letdown" of stable euthymia. The patient remained unsatisfied, however, and he decided to consciously increase his intake of caffeinated beverages (coffee, tea, soda) and over-the-counter caffeine tablets in the hope that he could recapture some of his lost vigor. Within 3 days, the patient did indeed feel more energetic and less "blunted." Encouraged by this early result, the patient adhered to this unilateral plan of high caffeine intake. After another week had passed, he had so much energy that he needed only 2–3 hours of sleep each night to feel completely refreshed. He felt so good that he did not even mind taking the lithium, which he continued to do faithfully. Two weeks later, he felt he was doing so well that he was going to quit his job as a cook and travel to Paris to open his own bistro. He did not speak any French, but that seemed only a minor impediment. At that point, the patient had one of his scheduled appointments with his psychiatrist. He was tempted to skip it but decided to go and say goodbye before his flight the next day. The patient ebulliently shared his future plans with the psychiatrist, who artfully convinced the patient to defer his plans and accompany him to the day hospital, where a lithium blood level was found to be only 0.3 mEq/L.

Discussion

This is an example of the role of caffeine in decreasing lithium blood levels.[1]

Lithium is primarily renally excreted. Caffeine, as well as other xanthines such as theophylline, acts as a diuretic by increasing the glomerular filtration rate and renal blood flow, which results in increased excretion of most solutes, including sodium and lithium. This increased excretion of lithium led to a decrease in the lithium blood level and the recurrence of a manic episode of moderate to severe intensity. It is worth noting that different diuretic agents have different effects on lithium excretion. Thiazide diuretics increase lithium levels (see "DI Difficulties," p. 211) whereas xanthines, carbonic anhydrase inhibitors, and other osmotic diuretics will reliably decrease lithium levels (Kaplan and Sadock 1998).

References

Gray GE, Gray LK: Nutritional aspects of psychiatric disorders. J Am Diet Assoc 89(10):1492–1498, 1989

Kaplan H, Sadock B: Kaplan and Sadock's Synopsis of Psychiatry: Behavioral Sciences/Clinical Psychiatry, 8th Edition. Baltimore, MD, Williams & Wilkins, 1998, p 1052

---

[1]Gray and Gray 1989; Kaplan and Sadock 1998.

Decaffeination Intoxication

A 29-year-old woman with recurrent major depression had been responding well to venlafaxine (Effexor), 300 mg/day, augmented with lithium, 1,350 mg qhs (blood level=0.85 mEq/L). However, she had also been reporting chronic sleep difficulties, which had not reliably responded to trazodone (Desyrel), low-dose mirtazapine (Remeron), or various sedative-hypnotic agents. She visited a sleep specialist (neurologist), who took a full sleep hygiene and dietary inventory and discovered that the patient was a prolific consumer of coffee and caffeinated soda throughout the day. He advised the patient to gradually decrease her intake of caffeine in the hope that this would prove more helpful than aggressive sleep-inducing medications. Within 2 weeks, the patient had successfully weaned herself from all caffeinated beverages. While she had expected some transient fatigue in the caffeine-tapering period, her malaise and significant sedation were well beyond anything she had anticipated. She even became tremulous and slightly confused, but she attributed these symptoms to withdrawal from the caffeine and regarded them as a sign of just how physiologically dependent she had become on caffeine. When a full week free of caffeine did not produce any improvements, however, she suspected another cause for her difficulties. She contacted her psychiatrist and described the findings and recommendations of the sleep specialist. After discussing options, the psychiatrist opted to allow the patient to remain at home, but he instructed her to skip her next dose of lithium and then reduce her dose to 900 mg qhs. One week later, her blood level was 0.9 mEq/L at a dosage of only 900 mg/day of lithium.

Discussion

This is an example of a state of mild lithium toxicity resulting from the discontinuation of caffeine.[2]

As stated in the previous case ("More Than He Bargained For"), lithium is primarily renally excreted. Caffeine, as well as other xanthines like theophylline, acts as a diuretic by increasing the glomerular filtration rate and renal blood flow, which results in increased excretion of most solutes, including sodium and lithium. The patient's baseline lithium dosage (1,350 mg/day) was able to generate the lithium blood level of 0.85

---

[2]Gray and Gray 1989; Kaplan and Sadock 1998.

mEq/L even in the face of caffeine's excretion of lithium. When the caffeine was removed, however, the rate of lithium excretion declined, leading to an increase in the lithium blood level and symptoms of mild lithium toxicity at the lithium dosage of 1,350 mg/day. The psychiatrist compensated for this likely state of mild lithium toxicity by decreasing the lithium dosage by 33% (to a new dosage of 900 mg/day). Even with this decrease in dosage, the patient's "decaffeinated" lithium level increased slightly (from 0.85 mEq/L to 0.9 mEq/L).

References

Gray GE, Gray LK: Nutritional aspects of psychiatric disorders. J Am Diet Assoc 89(10):1492–1498, 1989

Kaplan H, Sadock B: Kaplan and Sadock's Synopsis of Psychiatry: Behavioral Sciences/Clinical Psychiatry, 8th Edition. Baltimore, MD, Williams & Wilkins, 1998, p 1052

The Tremulous Trucker

A 49-year-old truck driver with bipolar I disorder had maintained good clinical stability while taking risperidone (Risperdal), 2 mg qhs, and lithium, 900 mg/day (mean blood level=0.9 mEq/L) for the past 3 years. During an appointment with his internist, the patient's blood pressure was found to be 175/100 mm Hg. His internist prescribed valsartan (Diovan), 80 mg/day, and asked him to return in 3 months for a follow-up appointment. Three days later, the patient noticed that his hands were oddly tremulous, requiring him to grip the steering wheel firmly with both hands to ensure that he retained control of his vehicle. The tremors grew worse over the ensuing 4 days, as did his fatigue and mild confusion. When he woke up the next morning, he had trouble maintaining his balance while trying to walk. He then reported his difficulties to his internist, who advised him to report to the nearest ER, where his lithium level was found to be 1.4 mEq/L ("Lithium/Valsartan Possible Interaction" 2000).

Discussion

This is an example of mild lithium toxicity caused by valsartan, a medication that inhibits the production of aldosterone.[3]

---

[3]"Lithium/Valsartan Possible Interaction" 2000.

Lithium is excreted by the kidney. Valsartan is an angiotensin II receptor antagonist. Under normal circumstances, circulating angiotensin II binds to specific receptors, which then trigger aldosterone release. Angiotensin II receptor antagonists, such as valsartan, inhibit the ability of angiotensin II to stimulate aldosterone release. Since aldosterone is a hormone that promotes the retention of sodium and the excretion of potassium, the inhibition of aldosterone release leads to increased sodium excretion and potassium retention. Sodium and lithium compete for reabsorption at the proximal tubule of the nephron. With relatively less sodium being retained because of angiotensin II receptor antagonist–induced decreases in aldosterone release, lithium is better able to compete for reabsorption, leading to increased lithium retention and possible toxicity. In this case, the lithium level increased by roughly 50%.

References

Lithium/valsartan possible interaction. Psychiatry Drug Alerts 14(7):55, 2000. Available at: http://www.alertpubs.com/Issues/2000/Sda72000.PDF. Accessed August 21, 2002

Over-the-Counter Calamity

A 44-year-old woman with a long-standing atypical depression had done well for the past 2 years since she began taking tranylcypromine (Parnate), 60 mg/day. One winter, she contracted a bad case of the flu. Her rhinitis, congestion, and sore throat made it difficult for her to sleep. In an exhausted and inattentive state, she purchased some NyQuil and ingested a dose before going to bed. Instead of getting a good night's sleep, she became acutely anxious and agitated, diaphoretic, febrile (again), nauseated, and "twitchy," as she described it. Fortunately, these symptoms basically remitted by the morning, and when she read the ingredients on the NyQuil box she realized her mistake.

Discussion

This is an example of a substrate added to an inhibitor.

In this case, a mild central serotonin syndrome arose from the combination of tranylcypromine, a monoamine oxidase inhibitor (MAOI), and dextromethorphan (contained in NyQuil), which has "serotomimetic" properties. When the dextromethorphan increased the availability of serotonin (a monoamine), this synergized with the prevention of the breakdown of serotonin, which defines one of the functions of an MAOI. This combination of factors increased physiologic serotonin tone to the extent that a central serotonin syndrome was produced (Bem and Peck 1992). In this case, which was relatively mild, hospitalization was not necessary, but safe clinical practice entails a low threshold for mobilizing intensive medical interventions when this syndrome is suspected.

References

Bem JL, Peck R: Dextromethorphan: an overview of safety issues. Drug Saf 7(3): 190–199, 1992

Mood Destabilization

A 29-year-old woman with rapid cycling bipolar I disorder had responded well to lamotrigine (Lamictal) monotherapy at a dosage of 250 mg/day. She unfortunately required hospitalization for a breakthrough manic episode. She had a history of poorly tolerating lithium and antipsychotic medications (even atypical agents). Since both divalproex sodium (Depakote) and carbamazepine (Tegretol) are notorious for altering lamotrigine blood levels, her psychiatrist decided to try oxcarbazepine (Trileptal), having heard that it was an effective antimanic agent and that it reportedly did not interact with other medications as carbamazepine did. He had the patient start taking oxcarbazepine, 300 mg bid, and titrated the dosage to 600 mg bid; he also provided lorazepam (Ativan) for control of agitation. The patient's manic symptoms gradually improved, and she was discharged after a 10-day stay in the hospital. However, 3 weeks later, the patient began to experience an emergence of depressive symptoms. Her psychiatrist responded by (gradually) increasing the dosage of lamotrigine to 400 mg/day, which did produce a remission of her depressive symptoms.

Discussion

This is an example of an inducer added to a substrate.

Lamotrigine is primarily metabolized through phase II glucuronidation, specifically by the 1A4 enzyme (Hiller et al. 1999). Although it is not the robust 3A4 inducer that carbamazepine is, oxcarbazepine is certainly not devoid of any drug interactions. Oxcarbazepine is a moderately strong inducer of glucuronidation enzyme 1A4 (May et al. 1999). This led to an increase in the amount of this enzyme that was available to metabolize the lamotrigine, resulting in a decrease in the blood level of lamotrigine, even though the lamotrigine dosage had not been decreased. Studies have demonstrated that the addition of oxcarbazepine to lamotrigine can decrease lamotrigine blood levels by roughly 30% on average (May et al. 1999). Although this is not a huge effect, it accounted for lamotrigine's loss of mood stabilizer efficacy (for the depressive pole), and thus the patient's "mood instability" following the addition of oxcarbazepine, another "mood stabilizer." The psychiatrist nicely compensated for this effect by increasing the lamotrigine dosage from 250 mg/day to 400 mg/day.

References

Hiller A, Nguyen N, Strassburg CP, et al: Retigabine N-glucuronidation and its potential role in enterohepatic circulation. Drug Metab Dispos 27(5):605–612, 1999

May TW, Rambeck B, Jurgens U: Influence of oxcarbazepine and methsuximide on lamotrigine concentrations in epileptic patients with and without valproic acid comedication: results of a retrospective study. Ther Drug Monit 21(2): 175–181, 1999

Migraineur

A 30-year-old woman with a history of anxious depression was responding well to sertraline (Zoloft), 100 mg/day. Two months after starting a new job, she consulted with a neurologist for treatment of migraine headaches, which had grown more frequent and painful since she began her new job. The neurologist prescribed sumatriptan (Imitrex), 75 mg po prn for migraine headache. One week later, she had a migraine and took one of the sumatriptan tablets. Within 2 hours, she experienced myoclonic jerking of her extremities, flushing, fever, nausea, and tremor. She contacted her neurologist, who advised her to report to the ER immediately. She recovered fully after 48 hours of observation and treatment with lorazepam (Ativan) and antipyretic medication.

Discussion

This is an example of a central serotonin syndrome caused by the additive and possibly synergistic effects of sertraline and sumatriptan.[4]

---

[4]Imitrex (package insert) 2001.

Sertraline is a potent inhibitor of serotonin reuptake, which makes serotonin more systemically available. Sumatriptan is a potent and specific agonist of the serotonin receptor $5\text{-}HT_{1D}$. Sertraline nonspecifically mobilizes more serotonin to bind with the whole array of postsynaptic serotonin receptors, including $5\text{-}HT_{1D}$, via serotonin reuptake inhibition. This effect seems to have combined with sumatriptan's specific $5\text{-}HT_{1D}$ agonism, and possibly other postsynaptic serotonergic effects, to produce a central serotonin syndrome (Hojer et al. 2002) in this patient.

It is worth noting that this interaction between sertraline (or any of the SSRIs) and sumatriptan (or any of the "triptans") is quite uncommon. However, it is frequent enough and serious enough that sumatriptan should not be added to these agents in a routine, casual, or unreflective manner, as if it was a matter of standard practice. These combinations are certainly not absolutely contraindicated, but care should be exercised with sumatriptan dosing, and good communication and contingency planning in the event of a worst-case scenario should be undertaken prospectively with patients who are taking these medications concurrently.

References

Hojer J, Personne M, Skagius AS, et al: Serotonin syndrome: several cases of this often overlooked diagnosis. Lakartidningen 99(18):2054–2055, 2058–2060, 2002

Imitrex (package insert). Research Triangle Park, NC, GlaxoSmithKline, 2001

Stupor

A 57-year-old man had been receiving carbamazepine (Tegretol), 800 mg/day (blood level=10.7 μg/mL), for the successful treatment of trigeminal neuralgia. He had also been experiencing increasing heartburn and obvious gastroesophageal reflux. He reported these symptoms to his internist, who prescribed omeprazole (Prilosec), 20 mg/day, for the patient. One week later, the patient noted increasing fatigue, dizziness, and diplopia. His wife informed him that his eyes were "fluttering" at times. Over the next 4 days, the patient became increasingly confused and even frankly stuporous at times. The patient's wife finally brought him to the local ER, where his carbamazepine level was found to be 19.5 μg/mL (Dixit et al. 2001).

Discussion

This is an example of an inhibitor added to a substrate.

An examination of possible P-450-related factors that could impact on this interaction is a confusing enterprise. First, carbamazepine is primarily metabolized at 3A4, although 1A2 and 2C9 play more minor roles in carbamazepine's metabolism (Spina et al. 1996). Omeprazole is an inhibitor of 2C19 and an inducer of 1A2 (Furuta et al. 2001; Nousbaum et al. 1994). The sum of these factors suggests that the addition of omeprazole should lead to a modest decrease in the blood level of carbamazepine via induction of 1A2. There are no known relevant glucuronidation issues here.

However, carbamazepine is a substrate of the P-glycoprotein transporter (Potschka et al. 2001), and omeprazole is an inhibitor of the activity of this transporter (Pauli-Magnus et al. 2001). Thus, the addition of omeprazole significantly impaired the ability of the P-glycoprotein transporter to extrude carbamazepine from enterocytes back into the gut lumen. Since more carbamazepine was retained in enterocytes, there was greater bioavailability and absorption of carbamazepine from the gut, leading to an increase in the blood level of carbamazepine (Dixit et al. 2001). Carbamazepine blood level increases of 50%–100% with this combination are not unusual.

References

Dixit RK, Chawla AB, Kumar N, et al: Effect of omeprazole on the pharmacokinetics of sustained-release carbamazepine in healthy male volunteers. Methods Find Exp Clin Pharmacol 23(1):37–39, 2001

Furuta S, Kamada E, Suzuki T, et al: Inhibition of drug metabolism in human liver microsomes by nizatidine, cimetidine and omeprazole. Xenobiotica 31(1):1–10, 2001

Nousbaum JB, Berthou F, Carlhant D, et al: Four-week treatment with omeprazole increases the metabolism of caffeine. Am J Gastroenterol 89(3):371–375, 1994

Pauli-Magnus C, Rekersbrink S, Klotz U, et al: Interaction of omeprazole, lansoprazole and pantoprazole with P-glycoprotein. Naunyn Schmiedebergs Arch Pharmacol 364(6):551–557, 2001

Potschka H, Fedrowitz M, Loscher W: P-glycoprotein and multidrug resistance-associated protein are involved in the regulation of extracellular levels of the major antiepileptic drug carbamazepine in the brain. Neuroreport 12(16):3557–3560, 2001

Spina E, Pisani F, Perucca E: Clinically significant pharmacokinetic drug interactions with carbamazepine: an update. Clin Pharmacokinet 31(3):198–214, 1996

Bella Donna

A 35-year-old woman with panic disorder and a history of polysubstance dependence was taking citalopram (Celexa), 40 mg/day, and imipramine (Tofranil), 150 mg/day (imipramine+desipramine blood level= roughly 200 ng/mL). One winter, she contracted a common cold with significant nasal congestion and sinus pain, and she opted to take some over-the-counter diphenhydramine (Benadryl), 25 mg roughly three times per day. Within 3 days, she was experiencing blurry vision, dry mouth, and constipation. She reported her difficulties to her psychiatrist, who acknowledged her addiction concerns about pseudoephedrine-containing decongestants and advised her to take a nasal spray that contained oxymetazoline HCl for 2–3 days as needed. Her symptoms abated within 2 days of discontinuing the diphenhydramine.

Discussion

This is an example of excessive pharmacodynamic synergy.

Imipramine is a strongly anticholinergic compound, which means that it avidly blocks muscarinic receptors. Diphenhydramine, a nonspecific antihistamine, is also a strongly anticholinergic compound (Radovanovic et al. 2000). When combined in this individual, they produced several symptoms (blurry vision, dry mouth, constipation) that can emerge when muscarinic receptors are blocked too robustly. Other classic anticholinergic symptoms include tachycardia, urinary retention, and memory disturbance, confusion, and delirium (Gershon 1984).

The anticholinergic symptom of blurry vision is caused by pupillary dilatation (mydriasis). Legend has it that in order to enhance their attractiveness, women were encouraged to consume belladonna (which in Italian means "beautiful lady"), a strongly anticholinergic plant, to produce pupillary dilatation.

References

Gershon S: Comparative side effect profiles of trazodone and imipramine: special reference to the geriatric population. Psychopathology 17 (suppl 2):39–50, 1984

Radovanovic D, Meier PJ, Guirguis M, et al: Dose-dependent toxicity of diphenhydramine overdose. Hum Exp Toxicol 19(9):489–495, 2000

A Risky Regimen

A hospitalized 28-year-old man with treatment-resistant, rapid-cycling bipolar I disorder and a comorbid seizure disorder had been consistently seizure-free, but his bipolar illness had not responded to the following regimen: lithium, 1,050 mg/day (blood level=1.0 mEq/L); divalproex sodium (Depakote), 1,500 mg/day (blood level=120 μg/mL); clonazepam (Klonopin), 2 mg bid; loxapine (Loxitane), 50 mg/day; and benztropine, 1 mg bid. His psychiatrist decided to add carbamazepine and thus go to "triple therapy." Over a 2-week period, the carbamazepine was titrated to a dosage of 800 mg/day (blood level=8.7 μg/mL). After 2 weeks at this dosage of carbamazepine, the patient was increasingly lethargic and irritable, and the staff observed that his gait had become unsteady. He reported the emergence of auditory hallucinations and seemed preoccupied with the observation cameras on the unit. The next morning, many laboratory studies were obtained. Later that day, he had his first grand mal seizure in 2 years. Significant test results included: divalproex level=48 μg/mL, carbamazepine level=9.2 μg/mL, ALT=82 IU/L (previously WNL), and ammonia level=65 μmoles/L.

Discussion

This is a mixed example of an inducer (carbamazepine) added to a substrate (divalproex) and a substrate (carbamazepine) added to an inhibitor (divalproex).

First, the metabolism of divalproex is exceedingly complex, involving multiple P450 enzymes and phase II glucuronidation, and liberating 50 or more metabolites (Pisani 1992)! Blood levels of free valproate are also influenced by the presence of other strongly plasma protein–bound compounds. Carbamazepine is an inducer of 1A2, 3A4, and glucuronidation enzyme 1A4 (Ketter et al. 1999; Parker et al. 1998; Rambeck et al. 1996; Spina et al. 1996). Of these possible points of interaction, the most likely means by which carbamazepine influences the metabolism of divalproex is through the increased production (induction) of glucuronidation enzymes, leading to a roughly 60% decrease in valproate blood levels (Jann et al. 1988). However, through P450 mechanisms that have not yet been fully elucidated, the coadministration of carbamazepine and divalproex leads to the increased production of a hepatotoxic metabolite, 4-ene-valproic acid. (The production of 4-ene-valproic acid is increased when divalproex is coadministered with agents that artificially induce its metabolism, such as phenytoin, phenobarbital, and carbamazepine [Levy et al.

1990].) Although such an increased metabolite level in the face of enzymatic induction may at first seem counterintuitive, it represents the enhanced generation of this toxic metabolite through the artificial "pulling" of divalproex's metabolism through atypical pathways at a greater than usual rate, which is faster than the body can eliminate it, thus leading to accumulation of this metabolite. The accumulation of this hepatotoxic 4-ene-valproic acid metabolite likely explains this patient's increased ALT and hyperammonemia, and therefore possibly some component of the patient's unsteadiness, lethargy, irritability, and/or seizure activity.

Second, carbamazepine is primarily a 3A4 substrate, although 1A2 and 2C9 also make minor contributions to carbamazepine's metabolism (Spina et al. 1996). Divalproex is an inhibitor of 2C9 (Wen et al. 2001) and glucuronidation enzyme 1A4 (Hachad et al. 2002). Divalproex does not usually strongly affect the metabolism of carbamazepine itself, except that through its 2C9 inhibition, it may contribute to modestly raising carbamazepine levels.

The main divalproex effect of relevance to this case, however, is divalproex's ability to inhibit epoxide hydrolase, the enzyme responsible for metabolizing and clearing the neurotoxic and pharmacologically active carbamazepine metabolite carbamazepine-10,11-epoxide (Spina et al. 1996). Inhibition of epoxide hydrolase by divalproex leads to the accumulation of the carbamazepine-10,11-epoxide metabolite. This metabolite may lead to clinical carbamazepine toxicity both because of its intrinsic neurotoxicity and because of the additive effects on the nervous system exerted by its carbamazepine-like pharmacologic activity. Thus, the formation of this metabolite, when considered in addition to the carbamazepine, likely played a large role in the patient's ataxia, emerging paranoia, lethargy, and seizure. Blood levels of carbamazepine-10,11-epoxide will often double when valproic acid is added (G. Krause, personal communication, April 2002). It is important to note that this state of clinical carbamazepine toxicity, due to the increased production of carbamazepine-10,11-epoxide, is usually accompanied by carbamazepine blood levels that are in the normal range. Laboratories do not typically test for the presence of either 4-ene-valproic acid or carbamazepine-10,11-epoxide. These tests must be specifically requested.

References

Hachad H, Ragueneau-Majlessi I, Levy RH: New antiepileptic drugs: review on drug interactions. Ther Drug Monit 24(1):91–103, 2002

Jann MW, Fidone GS, Israel MK, et al: Increased valproate serum concentrations upon carbamazepine cessation. Epilepsia 29(5):578–581, 1988

Ketter TA, Frye MA, Cora-Locatelli G, et al: Metabolism and excretion of mood stabilizers and new anticonvulsants. Cell Mol Neurobiol 19(4):511–532, 1999

Levy RH, Rettenmeier AW, Anderson GD, et al: Effects of polytherapy with phenytoin, carbamazepine, and stiripentol on formation of 4-ene-valproate, a hepatotoxic metabolite of valproic acid. Clin Pharmacol Ther 48(3):225–235, 1990

Parker AC, Pritchard P, Preston T, et al: Induction of CYP1A2 activity by carbamazepine in children using the caffeine breath test. Br J Clin Pharmacol 45(2): 176–178, 1998

Pisani F: Influence of co-medication on the metabolism of valproate. Pharmaceutisch Weekblad (Scientific Edition) 14(3A):108–113, 1992

Rambeck B, Specht U, Wolf P: Pharmacokinetic interactions of the new antiepileptic drugs. Clin Pharmacokinet 31(4):309–324, 1996

Spina E, Pisani F, Perucca E: Clinically significant pharmacokinetic drug interactions with carbamazepine: an update. Clin Pharmacokinet 31(3):198–214, 1996

Wen X, Wang JS, Kivisto KT, et al: In vitro evaluation of valproic acid as an inhibitor of human cytochrome P450 isoforms: preferential inhibition of cytochrome P450 2C9 (CYP2C9). Br J Clin Pharmacol 52(5):547–553, 2001

Mother Superior

A 75-year-old woman who served as the mother superior at a local convent had been diagnosed with bipolar II disorder. For most of her adult life, her symptoms had been poorly responsive to the range of antidepressants and mood stabilizers. They were either poorly tolerated or ineffective. Two years ago, however, her psychiatrist initiated a trial of lamotrigine (Lamictal), titrated to a dosage of 150 mg/day, to which she responded better than anyone had expected. However, she developed trigeminal neuralgia and visited a neurologist for evaluation and treatment. He prescribed carbamazepine (Tegretol), titrated to a dosage of 600 mg/day (blood level=7.5 μg/mL), which proved quite effective for her pain. (She had tolerated carbamazepine in the past, but it had proven ineffective for her mood-related symptoms.) One month after starting the carbamazepine, however, she began to experience the familiar and dreaded sensation of evolving depression. She quickly contacted her psychiatrist and informed him of her current clinical state, as well as the recent addition of carbamazepine to her regimen. The psychiatrist promptly gave her a titration schedule for a new lamotrigine target dosage of 300 mg/day. Within 3 weeks of achieving this new dosage, her depressive symptoms had remitted.

Discussion

This is an example of an inducer added to a substrate.

Lamotrigine is primarily metabolized through phase II glucuronidation, specifically by the uridine 5′-diphosphate glucuronosyltransferase (UGT) 1A4 enzyme (Hiller et al. 1999). Carbamazepine is an inducer of 1A2, 3A4, and glucuronidation enzyme 1A4 (Ketter et al. 1999; Parker et al. 1998; Rambeck et al. 1996; Spina et al. 1996). The addition of carbamazepine led to the increased production of 1A4, which was therefore able to metabolize the lamotrigine more efficiently, leading to a decrease in the blood level of lamotrigine, even though the lamotrigine dosage had not been decreased (Hachad et al. 2002; DeVane and Nemeroff 2002). This decrease in the lamotrigine blood level likely led to the patient's impending depression. The psychiatrist was aware that the presence of carbamazepine leads to significant reductions in lamotrigine blood levels. Accordingly, he titrated the lamotrigine dosage to double its previous amount (from 150 mg/day to 300 mg/day) and thus compensated for the effect of carbamazepine on the metabolism of lamotrigine.

There is some debate as to whether the addition of lamotrigine to carbamazepine increases the production of the neurotoxic carbamazepine-10,11-epoxide metabolite. A 1992 study suggested that this was the case (Warner et al. 1992), yet several other studies from 1997 onward strongly suggested that there is no increase in carbamazepine-10,11-epoxide with the addition of lamotrigine (Besag et al. 1998; Eriksson and Boreus 1997; Gidal et al. 1997). A prudent course would be to coadminister these agents as is clinically indicated but also to maintain an awareness of the possibility of such an interaction should symptoms of confusion and/or lethargy arise.

References

Besag FM, Berry DJ, Pool F, et al: Carbamazepine toxicity with lamotrigine: pharmacokinetic or pharmacodynamic interaction? Epilepsia 39(2):183–187, 1998

DeVane CL, Nemeroff CB: 2002 guide to psychotropic drug interactions. Primary Psychiatry 9(3):28–57, 2002

Eriksson AS, Boreus LO: No increase in carbamazepine-10,11-epoxide during addition of lamotrigine treatment in children. Ther Drug Monit 19(5):499–501, 1997

Gidal BE, Rutecki P, Shaw R, et al: Effect of lamotrigine on carbamazepine epoxide/carbamazepine serum concentration ratios in adult patients with epilepsy. Epilepsy Res 28(3):207–211, 1997

Hachad H, Ragueneau-Majlessi I, Levy RH: New antiepileptic drugs: review on drug interactions. Ther Drug Monit 24(1):91–103, 2002

Hiller A, Nguyen N, Strassburg CP, et al: Retigabine N-glucuronidation and its potential role in enterohepatic circulation. Drug Metab Dispos 27(5):605–612, 1999

Ketter TA, Frye MA, Cora-Locatelli G, et al: Metabolism and excretion of mood stabilizers and new anticonvulsants. Cell Mol Neurobiol 19(4):511–532, 1999

Parker AC, Pritchard P, Preston T, et al: Induction of CYP1A2 activity by carbamazepine in children using the caffeine breath test. Br J Clin Pharmacol 45(2): 176–178, 1998

Rambeck B, Specht U, Wolf P: Pharmacokinetic interactions of the new antiepileptic drugs. Clin Pharmacokinet 31(4):309–324, 1996

Spina E, Pisani F, Perucca E: Clinically significant pharmacokinetic drug interactions with carbamazepine: an update. Clin Pharmacokinet 31(3):198–214, 1996

Warner T, Patsalos PN, Prevett M, et al: Lamotrigine-induced carbamazepine toxicity: an interaction with carbamazepine-10,11-epoxide. Epilepsy Res 11(2): 147–150, 1992

Displaced

A 48-year-old man with a history of bipolar I disorder was responding well to treatment with divalproex sodium (Depakote), 1,250 mg/day (blood level=85 μg/mL), and quetiapine (Seroquel), 500 mg/day. One morning, he awoke to the sensation of mild tingling in his right hand and lower arm. When this persisted for more than 30 minutes, the patient contacted his internist, who advised the patient to report to the local ER for an evaluation. This symptom spontaneously abated after 4 hours, and there was some concern that this represented a transient ischemic attack (TIA), given the patient's as yet untreated hypertension (155/95 mm Hg) and family history that was positive for stroke. He was given enalapril (Vasotec), 5 mg bid, and aspirin, 325 mg/day. Within 3 days, he reported sudden onset of terrific fatigue and sedation, as well as incoordination. Fearing a cerebrovascular accident (CVA), the internist again recommended that the patient visit the ER. A consulting neurologist commented that his presentation was consistent with valproate toxicity, yet the divalproex blood level was essentially unchanged at 95 μg/mL.

Discussion

This is an example of elevation of the free fraction of a drug through displacement from plasma protein binding sites.

Divalproex is tightly bound to plasma proteins and thus may displace and be displaced by other tightly bound compounds, such as phenytoin and warfarin (Depakote [package insert] 2002). Aspirin is also highly plasma protein–bound, and the addition of aspirin to divalproex leads to active competition between these two agents for plasma protein binding sites. Thus, the aspirin displaced a significant portion of the plasma protein–bound (or inactive) divalproex into the free, unbound, or active fraction. Studies have demonstrated a fourfold increase in the free valproate fraction following the introduction of standing doses of aspirin ("Divalproex—RxList Monographs" 2002). This marked increase in the free/active fraction led to the development of marked sedation and incoordination. It is worth noting that the aspirin induced a shift in the ratio of bound/inactive divalproex to unbound/active divalproex, but the total sum of these two divalproex fractions remained essentially unchanged. This explains why the blood level was of little help in making this diagnosis. Standard divalproex blood levels measure the total of the bound and free portions. However, free valproate levels can be ordered and obtained, although these must be specifically requested as such, which would be prudent when combining valproate with other highly plasma protein–bound compounds.

This process also happens in the other direction. When aspirin is added to divalproex, less of the aspirin is bound to plasma proteins than would have otherwise been expected, thus yielding a higher free/active fraction of aspirin than would be typical at that dosage. For aspirin, however, this is seldom a clinically significant issue.

References

Depakote (package insert). North Chicago, IL, Abbott Laboratories, 2002

Divalproex—RxList Monographs, 2002. Available at: http://www.rxlist.com/cgi/generic/dival_ad.htm. Accessed August 21, 2002

High–Low

A 26-year-old man with bipolar I disorder and a seizure disorder was released from the hospital following successful treatment of a manic episode with a new regimen: divalproex sodium (Depakote), 1,500 mg/day (blood level=105 μg/mL), and olanzapine (Zyprexa), 20 mg/day. One week after discharge, his sister presented him with the phenytoin (Dilantin), 400 mg/day, that had been prescribed for his seizure disorder prior to his recent hospitalization. Without contacting his psychiatrist or neurologist, the patient again began taking the phenytoin. At 400 mg/day, his typical phenytoin blood level was roughly 17 μg/mL. Five days after resuming the phenytoin, he seemed more sedated than he had expected, but this was not so severe as to be troublesome. However, his sister did note that his "eyeballs seemed to twitch" at times. Two weeks later, the patient was becoming manic again, requiring less sleep and hatching grandiose and expensive plans. His sister contacted the psychiatrist, who met with the patient, discovered that he had resumed his phenytoin, and promptly advised him to stop this medication. He also increased the dosages of divalproex and olanzapine to 2,000 mg/day and 30 mg/day, respectively, and instructed the patient to take these medications at these dosages for the next 2 weeks, or until the patient developed side effects from these changes. Phenytoin and divalproex blood levels were also obtained, with results of 23 μg/mL and 62 μg/mL, respectively. These 2 weeks progressed uneventfully. The patient's manic symptoms gradually receded, and he then returned to his baseline dosages of divalproex and olanzapine without incident. His divalproex blood level also returned to its baseline.

Discussion

This is a complex example of mixed plasma protein displacement, P450 inhibition, and induction of glucuronidation.

First, the metabolism of divalproex is exceedingly complex, involving multiple P450 enzymes and phase II glucuronidation, and liberating 50 or more metabolites (Pisani 1992)! Blood levels of free valproate are also influenced by the presence of other strongly plasma protein–bound compounds, such as phenytoin. Also, olanzapine is a substrate of 1A2, 2D6, and glucuronidation enzyme 1A4 (Callaghan et al. 1999; Linnet 2002). Phenytoin is an inducer of 3A4, 2C9, 2C19, and glucuronidation enzyme 1A4 (Chetty et al. 1998; Hachad et al. 2002; Spina and Perucca 2002). Therefore, adding the phenytoin led to the increased production of 3A4, 2C9, 2C19, and glucuronidation enzyme 1A4, thus yielding more efficient metabolism of both the divalproex and the olanzapine. This led to a roughly 50% decrease in the measured total divalproex blood level (Pisani 1992) and a decrease in the blood level of olanzapine. (Despite phenytoin's displacement of bound valproate to the free fraction, the inductive influences of treatment with phenytoin predominate to produce a significant net decrease in divalproex levels.)

Second, phenytoin is a substrate primarily of 2C9 and 2C19. Divalproex is an inhibitor of 2C9 and an inhibitor of glucuronidation enzyme

1A4 (Hachad et al. 2002; Wen et al. 2001). Phenytoin and divalproex are also both highly bound to plasma proteins. Thus, there is a rapid increase in the total phenytoin blood level by virtue of 2C9 inhibition and plasma protein displacement, yielding a roughly 60% increase in the free/unbound fraction of phenytoin ("Divalproex—RxList Monographs" 2002), which is the pharmacologically active fraction.

References

Callaghan JT, Bergstrom RF, Ptak LR, et al: Olanzapine: pharmacokinetic and pharmacodynamic profile. Clin Pharmacokinet 37(3):177–193, 1999

Chetty M, Miller R, Seymour MA: Phenytoin autoinduction. Ther Drug Monit 20(1):60–62, 1998

Divalproex—RxList Monographs, 2002. Available at: http://www.rxlist.com/cgi/generic/dival_ad.htm. Accessed August 21, 2002

Hachad H, Ragueneau-Majlessi I, Levy RH: New antiepileptic drugs: review on drug interactions. Ther Drug Monit 24(1):91–103, 2002

Linnet K: Glucuronidation of olanzapine by cDNA-expressed human UDP-glucuronosyltransferases and human liver microsomes. Hum Psychopharmacol 17(5):233–238, 2002

Pisani F: Influence of co-medication on the metabolism of valproate. Pharmaceutisch Weekblad (Scientific Edition) 14(3A):108–113, 1992

Spina E, Perucca E: Clinical significance of pharmacokinetic interactions between antiepileptic and psychotropic drugs. Epilepsia 43 (suppl 2):37–44, 2002

Wen X, Wang JS, Kivisto KT, et al: In vitro evaluation of valproic acid as an inhibitor of human cytochrome P450 isoforms: preferential inhibition of cytochrome P450 2C9 (CYP2C9). Br J Clin Pharmacol 52(5):547–553, 2001

Seized by Sadness

A 25-year-old woman with bipolar I disorder and a seizure disorder had been successfully maintained on lamotrigine (Lamictal) monotherapy (300 mg/day) for 18 months, during which she had been moderately euthymic and virtually seizure-free. However, about 2 months ago, she experienced an increase in the frequency of her seizures from once every 6 months to once each week. Her neurologist could not pinpoint an obvious cause for this change in her seizure frequency, but he decided to add phenytoin, 300 mg/day (blood level=13.2 μg/mL), to her lamotrigine. She immediately stopped having weekly seizures. One month later, she reported recurring depressive symptoms to her psychiatrist, who learned only at that time about the addition of the phenytoin to her regimen. The psychiatrist instructed the patient to increase her lamotrigine

ficulties with premenstrual mood swings and irritability. She consulted her gynecologist, and not her psychiatrist, about these problems. Her gynecologist diagnosed her with premenstrual dysphoric disorder and prescribed a medication called Sarafem (fluoxetine), 20 mg/day. The patient was certainly alert to avoid any medications like Prozac, Zoloft, Paxil, Luvox, or Celexa, but since this seemed different she was not concerned. Five days into her treatment with Sarafem, she experienced fever, myoclonus, confusion, nausea, and vomiting. Her brother brought her to the ER, where she was admitted and treated for a severe central serotonin syndrome.

Discussion

This is an example of a "substrate" added to an inhibitor.

Fluoxetine, even under the trade name Sarafem, is a potent SSRI. Phenelzine is a monoamine oxidase inhibitor (MAOI), and as such it inhibits the breakdown of all monoamines (serotonin, norepinephrine, epinephrine, and dopamine). Since all SSRIs greatly increase the availability of serotonin, the combination of these effects led to a severe state of excessive serotonergic function, the central serotonin syndrome (Beasley et al. 1993). One would expect the same syndrome to arise if buspirone (BuSpar) was added to an MAOI (Hyman et al. 1995). Fortunately, with close observation, antipyretics, and the liberal use of lorazepam (Ativan), this patient's condition stabilized after 1 week, although fatal outcomes have occurred with this combination of medications (Beasley et al. 1993).

References

Beasley CM Jr, Masica DN, Heiligenstein JH, et al: Possible monoamine oxidase inhibitor–serotonin uptake inhibitor interaction: fluoxetine clinical data and preclinical findings. J Clin Psychopharmacol 13(5):312–320, 1993

Hyman S, Arana G, Rosenbaum J: Handbook of Psychiatric Drug Therapy, 3rd Edition. Boston, MA, Little, Brown, 1995, p 84

Orthopedic Recovery and Manic Recurrence

A 27-year-old squash player with bipolar I disorder was being treated with divalproex sodium (Depakote), 1,250 mg/day (blood level=81 μg/mL), when he also began taking chronic indomethacin (Indocin), 50 mg tid, for persistent tendinitis. After 1 month on this regimen, the patient's liver function tests (GGT [γ-glutamyltransferase] and ALT [alanine transaminase]) rose to more than double their baseline values. Although this need not have necessitated the discontinuation of divalproex, the psychiatrist and patient both felt more comfortable transitioning from divalproex to lithium. The patient eventually had his dosage of lithium titrated to 750 mg/day, with a blood level of 0.9 mEq/L, and his divalproex was tapered and discontinued. Three months later, the patient's tendonitis had resolved and his orthopedist felt that he should taper and discontinue the patient's indomethacin. Four weeks after the indomethacin had been discontinued, the patient began to experience the emergence of manic symptoms. His psychiatrist had a lithium level checked immediately and to his surprise found that it was only 0.4 mEq/L, although the patient reported compliance with the lithium. Although he did not understand why the lithium level had declined, he nonetheless increased the lithium dosage to 1,200 mg/day and added risperidone (Risperdal), titrated rapidly to a dosage of 4 mg/day. The patient's new lithium level was 0.9 mEq/L. He rapidly responded to these interventions, and euthymia was restored.

Discussion

This is an example of reversal of inhibition.

As mentioned in "The Tremulous Triathlete" (p. 207), lithium is entirely renally excreted, and therefore a variety of medications that alter renal function can alter the rate of lithium excretion. All NSAIDs (including indomethacin), with the exception of aspirin and sulindac (Clinoril), will often elevate lithium blood levels (Ragheb 1990). The change in lithium blood level can be quite variable and difficult to predict, so close observation and frequent lithium levels are warranted. The proposed mechanism for this interaction is that NSAIDs inhibit prostaglandin synthesis in the kidney, thus interfering with the excretion of lithium (Imbs et al. 1997). The indomethacin was already present when the titration to a therapeutic dose and blood level of lithium occurred. Therefore, the lithium blood level of 0.9 mEq/L, at a dosage of 750 mg/day, was higher than it would have been had the indomethacin not been present. Therefore, with the discontinuation of the indomethacin, the patient's kidneys were able to more efficiently excrete the lithium, leading to a decrease in the lithium level to 0.4 mEq/L and a subsequent vulnerability to a manic switch. The psychiatrist eventually compensated for this effect by increasing the lithium dosage to 1,200 mg/day, and the addition of risperidone also helped to restabilize the patient.

References

Imbs JL, Barthelmebs M, Danion JM, et al: Mechanisms of drug interactions with renal elimination of lithium (in French). Bull Acad Natl Med 181(4):685–697, 1997

Ragheb M: The clinical significance of lithium–nonsteroidal anti-inflammatory drug interactions. J Clin Psychopharmacol 10(5):350–354, 1990

The Zebras Are Loose

A 46-year-old man with gastroesophageal reflux disease was taking metoclopramide (Reglan), 15 mg qid, with good success. Following a heated divorce, he became acutely depressed and was prescribed venlafaxine (Effexor), which was titrated up to a dosage of 225 mg/day. After 5 days at that dosage, the patient acutely developed flushing, diaphoresis, fevers, myoclonus, vomiting, confusion, and irritability. He contacted his new psychiatrist, who advised him to report to the nearest ER, where he was diagnosed with a central serotonin syndrome.

Discussion

This is an example of excessive pharmacodynamic synergy, which produced a central serotonin syndrome.

Venlafaxine is a serotonin reuptake inhibitor (an SRI at low doses, not an SSRI), which leads to increased availability of serotonin and thus serotonergic activity. Metoclopramide is generally considered primarily as a dopamine blocker, but it is also a 5-HT_4 agonist (Sommers et al. 1996). This selective serotonergic agonism synergized with the activity of the serotonin reuptake inhibitor (venlafaxine) to produce serotonergic activity so excessive that it reached a toxic level in the form of a central serotonin syndrome (Fisher and Davis 2002).

Unlike the case of combining sumatriptan and sertraline (see "Migraineur," p. 221), this interaction is so infrequent that it does not even merit any particular vigilance. This documented interaction is described here in order to facilitate recognition of this rare event should it ever arise, and to support the notion that we can never prejudge any potential interaction as impossible. A case of central serotonin syndrome has also been documented in an instance when metoclopramide was combined with sertraline (Fisher and Davis 2002).

References

Fisher AA, Davis MW: Serotonin syndrome caused by selective serotonin reuptake inhibitors–metoclopramide interaction. Ann Pharmacother 36(1):67–71, 2002

Sommers DK, Snyman JR, van Wyk M: Effects of metoclopramide and tropisetron on aldosterone secretion possibly due to agonism and antagonism at the 5-HT_4 receptor. Eur J Clin Pharmacol 50(5):371–373, 1996

Too Tired

A 52-year-old woman with generalized anxiety disorder and gastroesophageal disease had been taking alprazolam (Xanax), 1 mg tid, and cimetidine (Tagamet), 300 mg qid, for the past 3 months. Her anxiety had lessened, but the patient was still not satisfied with the results, and her psychiatrist was reluctant to further increase the dosage of alprazolam. The patient decided to take a "natural" remedy for her anxiety to augment the effect of the alprazolam. She purchased some kava *(Piper methysticum)* and began taking it per the pharmacist's instructions. Within 3 days, the patient had become so sedated that her brother could not arouse her from her slumber, at which point he called 911 and had the patient transported to the nearest ER. Several hours later, she became arousable and informed the ER staff about her use of kava (Almeida and Grimsley 1996).

Discussion

This is an example of excessive pharmacodynamic synergy resulting in excessive sedation, possibly worsened by a P450 inhibitor–substrate interaction.

Kava and alprazolam both act as agonists at the γ-aminobutyric acid (GABA) receptor, and they clearly act synergistically in this regard (Izzo and Ernst 2001). The magnitude of the resulting sedation in this case may well have been due to the presence of cimetidine. Cimetidine is an inhibitor of 3A4, 2D6, and 1A2 (Martinez et al. 1999). The cimetidine certainly impaired the ability of 3A4 to metabolize the alprazolam (Dresser et al. 2000), leading to a higher alprazolam blood level than would have been expected had the cimetidine not been present. Kava's metabolism has not been well characterized as yet, but it may be that the cimetidine also impaired the metabolism of kava as well as that of the alprazolam.

References

Almeida JC, Grimsley EW: Coma from the health food store: interaction between kava and alprazolam. Ann Intern Med 125(11):940–941, 1996

Dresser GK, Spence JD, Bailey DG: Pharmacokinetic-pharmacodynamic consequences and clinical relevance of cytochrome P450 3A4 inhibition. Clin Pharmacokinet 38(1):41–57, 2000

Izzo AA, Ernst E: Interactions between herbal medicines and prescribed drugs: a systematic review. Drugs 61(15):2163–2175, 2001

Martinez C, Albet C, Agundez JA, et al: Comparative in vitro and in vivo inhibition of cytochrome P450 CYP1A2, CYP2D6, and CYP3A by H2-receptor antagonists. Clin Pharmacol Ther 65(4):369–376, 1999

Kava Akinesia

A 75-year-old man with Parkinson's disease had been treated with carbidopa/levodopa (Sinemet) for the past 5 years, with reasonable control of his parkinsonian symptoms. His anxiety was increasing, however, in the face of increasing difficulty with daily tasks. His daughter convinced him to try a "natural remedy" called kava *(Piper methysticum).* He purchased some kava and took it as directed. Within 1 week, he experienced worsening immobility and akinesia. He informed his neurologist of this development and his recent use of kava. The neurologist advised discontinuing the kava, and within several days he had returned to his functional baseline.

Discussion

This is an example (I believe) of pharmacodynamic antagonism.

The levodopa component of Sinemet is clearly a dopamine agonist. Although this has not been well characterized to date, the addition of kava seemed to exert some sort of antagonistic effect on dopamine receptors in the basal ganglia, such that akinesia resulted. There have been several documented instances of this specific interaction; thus, it should not be regarded as unlikely (Izzo and Ernst 2001).

References

Izzo AA, Ernst E: Interactions between herbal medicines and prescribed drugs: a systematic review. Drugs 61(15):2163–2175, 2001

Natural Disaster (IV)

A 76-year-old man with a history of TIAs was taking aspirin (Ecotrin), 325 mg/day, as an anticoagulant agent. He began to experience a subjective sense of word-finding difficulty and slowed thinking. He feared that he was beginning to display symptoms of vascular dementia. He read in a popular magazine about the cognition-enhancing effects of ginkgo *(Ginkgo biloba)*, and he decided on his own to start taking ginkgo concentrated extract, 40 mg bid. One week later, the patient found two large, painful bruises and several small petechiae on his arms. He contacted his neurologist, who advised discontinuing the ginkgo. After stopping the ginkgo, the patient had no further hemorrhagic events (Rosenblatt and Mindel 1997).

Discussion

This is an example of excessive pharmacodynamic synergy, resulting in excessive inhibition of platelet aggregation.

Aspirin is used as a prophylactic agent to prevent ischemic vascular events by virtue of its ability to inhibit platelet aggregation. Ginkgo, in addition to actual demonstrated efficacy in the treatment of Alzheimer's dementia (Loew 2002), also shares this profile of inhibiting platelet aggregation. The antiplatelet effects of these two agents synergized to produce excessive anticoagulation. Cessation of the ginkgo reversed this effect (Lambrecht et al. 2000).

This type of synergy also occurs when warfarin (Coumadin) is combined with ginkgo (Lambrecht et al. 2000). There is a documented case of a 78-year-old woman who suffered a left parietal hemorrhagic CVA as a result of this interaction.

References

Lambrecht JE, Hamilton W, Rabinovich A: A review of herb-drug interactions: documented and theoretical. U.S. Pharmacist Home 25:8, August 2000. Available at: http://www.uspharmacist.com/NewLook/DisplayArticle.cfm?item_num=566. Accessed July 2, 2002

Loew D: Value of *Ginkgo biloba* in treatment of Alzheimer dementia. Wien Med Wochenschr 152(15–16):418–422, 2002

Rosenblatt M, Mindel J: Spontaneous hyphema associated with ingestion of *Ginkgo biloba* extract (letter). N Engl J Med 336(15):1108, 1997

Herbal Hemorrhage

A 35-year-old woman with rheumatic heart disease had been taking warfarin (Coumadin), 5 mg/day (International Normalized Ratio, or INR = 3.2), as an anticoagulant after her mitral valve replacement procedure. She was experiencing menstrual irregularity, so she visited an herbalist, who recommended danshen (*Salvia* spp.) for this problem. Within the next week, the woman was alarmed to discover she was experiencing spontaneous subcutaneous hematoma formation. She quickly reported to a local ER, where her INR was 9.8 and her PTT (partial thromboplastin time) was greater than 120 seconds (Yu et al. 1997).

Discussion

This is an example of pharmacodynamic synergy resulting in excessive anticoagulation.

Warfarin functions as an effective anticoagulant by virtue of its ability to inhibit specific steps of the coagulation cascade, as measured by the INR. Danshen is able to inhibit platelet aggregation (Lambrecht et al. 2000). These anticoagulant effects synergized to produce spontaneous hemorrhagic events, but fortunately not to a fatal extent in this case.

References

Lambrecht JE, Hamilton W, Rabinovich A: A review of herb-drug interactions: documented and theoretical. U.S. Pharmacist Home 25:8, August 2000. Available at: http://www.uspharmacist.com/NewLook/DisplayArticle.cfm?item_num=566. Accessed July 2, 2002

Yu CM, Chan JC, Sanderson JE: Chinese herbs and warfarin potentiation by 'danshen.' J Intern Med 241(4):337–339, 1997

TIA

A 65-year-old man with a history of atrial fibrillation with regular rate and prior cerebral vascular accidents (CVAs) was being maintained on warfarin (Coumadin), 2.5 mg/day (International Normalized Ratio, or INR= 2.6). A friend told him about ginseng *(Panax ginseng)* as a means of improving cognitive function and one's overall sense of well-being. He decided to purchase some ginseng and took it as directed by the herbalist. Three weeks later, he awoke with a tingling sensation over the lower left side of his face. He promptly called an ambulance and was transported to the nearest ER. He was found to be having a transient ischemic attack (TIA), and his INR was only 1.4 (Janetzky and Morreale 1997).

Discussion

This is an example of conflicting pharmacodynamic effects.

Warfarin functions as an effective anticoagulant by virtue of its ability to inhibit specific steps of the coagulation cascade, as measured by the INR. Ginseng appears to possess some procoagulant effects that decrease the INR, although the precise nature of these is not yet well understood (Lambrecht et al. 2000). The addition of ginseng to the warfarin antagonized some of warfarin's anticoagulant efficacy, making this patient vulnerable to clot formation, which might lead to TIAs or CVAs.

References

Janetzky K, Morreale AP: Probable interaction between warfarin and ginseng. Am J Health Syst Pharm 54(6):692–693, 1997

Lambrecht JE, Hamilton W, Rabinovich A: A review of herb-drug interactions: documented and theoretical. U.S. Pharmacist Home 25:8, August 2000. Available at: http://www.uspharmacist.com/NewLook/DisplayArticle.cfm?item_num=566. Accessed July 2, 2002

Fatal Error

A 53-year-old woman with type II diabetes mellitus and atypical major depression was being stably maintained on tranylcypromine (Parnate), 60 mg/day, and glipizide (Glucotrol), 5 mg/day. She began to experience some neuropathic pain in her feet, so her new internal medicine resident at her clinic started her on imipramine (Tofranil), 50 mg/day for 3 days and 100 mg/day thereafter. Within 1 week, she was experiencing a severe headache, tremors, vomiting, and myoclonus. Before she could seek help, she experienced a seizure and was incapacitated. She was eventually discovered by her daughter, who called 911. Once in the ER, she was found to have a blood pressure of 290/150 mm Hg, which caused a severe hemorrhagic CVA, and she was in the throes of a severe central serotonin syndrome. Efforts to resuscitate her were unsuccessful and she expired.

Discussion

This is an example of a “substrate” added to an inhibitor.

Tranylcypromine is a monoamine oxidase inhibitor (MAOI), so the breakdown of monoamines (serotonin, norepinephrine, epinephrine, and dopamine) was inhibited and these monoamines were more available and active throughout the body. Imipramine is a tricyclic antidepressant (TCA) with a mixed serotonergic and noradrenergic reuptake inhibition profile, which also caused these two neurotransmitters to be more available and active (Kaplan and Sadock 1998). The combination of these reuptake inhibition and monoamine oxidase inhibition influences led to a state of catecholamine and serotonin excess, which resulted in a hypertensive crisis superimposed on a severe central serotonin syndrome.

While this is a perilous combination of medications under the best of circumstances, it is even more dangerous when the TCA is added to a full dose of an MAOI, as opposed to the reverse sequence (Hyman et al. 1995). In this case, the results were fatal.

References

Hyman S, Arana G, Rosenbaum J: Handbook of Psychiatric Drug Therapy, 3rd Edition. Boston, MA, Little, Brown, 1995, p 74

Kaplan H, Sadock B: Kaplan and Sadock’s Synopsis of Psychiatry: Behavioral Sciences/Clinical Psychiatry, 8th Edition. Baltimore, MD, Williams & Wilkins, 1998, p 1103

Flagellated

A 25-year-old woman with bipolar I disorder was hospitalized because of a manic episode precipitated by noncompliance with her medications. Once admitted, she was restarted on carbamazepine (Tegretol), 800 mg/day (blood level=10.8 µg/mL), to which her manic symptoms were responding. While in the hospital, she complained about vaginal itching and burning, so a gynecological consultation was arranged for the patient. The gynecologist diagnosed her with bacterial vaginosis and prescribed metronidazole (Flagyl), 500 mg bid for 7 days. Two days later, the patient was notably ataxic on the unit, and she reported feeling dizzy and very sedated, as if intoxicated. A carbamazepine level drawn at that time was 27.0 µg/mL (L. Lin, K. Walters, personal communication, August 2002).

Discussion

This is (I believe) an example of a P-glycoprotein inhibitor added to a P-glycoprotein substrate.

Carbamazepine is primarily a 3A4 substrate (Spina et al. 1996), and metronidazole has been supposed to act as a 3A4 inhibitor in elevating carbamazepine blood levels. However, metronidazole does not raise the blood levels of either alprazolam (Blyden et al. 1988) or midazolam (Wang et al. 2000). It is difficult to understand how a 3A4 inhibitor capable of raising carbamazepine blood levels could fail to raise levels of triazolobenzodiazepines, which are more "selective" 3A4 substrates than carbamazepine (Dresser et al. 2000). Metronidazole has also been demonstrated to raise phenytoin blood levels (Blyden et al. 1988). Phenytoin is a substrate of 2C9 and 2C19 (Cadle et al. 1994; Mamiya et al. 1998). However, metronidazole fails to raise the blood levels of either tolbutamide (a 2C9 substrate) (Back and Tjia 1985; Jones et al. 1996) or diazepam (a 2C19 substrate) (Jensen and Gugler 1985; Ono et al. 1996). However, both carbamazepine and phenytoin are P-glycoprotein substrates (Potschka et al. 2001; Weiss et al. 2001). Also, metronidazole has been reported to raise the blood levels of both cyclosporine and tacrolimus blood levels (Herzig and Johnson 1999), and these are both P-glycoprotein substrates as well (Arima et al. 2001; Yokogawa et al. 2002).

From this evidence, I believe it is reasonable to regard metronidazole as a presumptive P-glycoprotein inhibitor. In that case, the metronidazole likely decreased the activity of P-glycoprotein, which led to less carbamazepine being extruded from enterocytes back into the gut lumen, where it would have been excreted rather than absorbed. The resulting increase in the absorption of carbamazepine probably explains the increase in the carbamazepine blood level following the addition of metronidazole.

References

Arima H, Yunomae K, Hirayama F, et al: Contribution of P-glycoprotein to the enhancing effects of dimethyl-beta-cyclodextrin on oral bioavailability of tacrolimus. J Pharmacol Exp Ther 297(2):547–555, 2001

Back DJ, Tjia JF: Inhibition of tolbutamide metabolism by substituted imidazole drugs in vivo: evidence for a structure-activity relationship. Br J Pharmacol 85(1): 121–126, 1985

Blyden GT, Scavone JM, Greenblatt DJ : Metronidazole impairs clearance of phenytoin but not of alprazolam or lorazepam. J Clin Pharmacol 28(3):240–245, 1988

Cadle RM, Zenon GJ 3rd, Rodriguez-Barradas MC, et al: Fluconazole-induced symptomatic phenytoin toxicity. Ann Pharmacother 28(2):191–195, 1994

Dresser GK, Spence JD, Bailey DG: Pharmacokinetic-pharmacodynamic consequences and clinical relevance of cytochrome P450 3A4 inhibition. Clin Pharmacokinet 38(1):41–57, 2000

Herzig K, Johnson DW: Marked elevation of blood cyclosporin and tacrolimus levels due to concurrent metronidazole therapy. Nephrol Dial Transplant 14(2): 521–523, 1999

Jensen JC, Gugler R: Interaction between metronidazole and drugs eliminated by oxidative metabolism. Clin Pharmacol Ther 37(4):407–410, 1985

Jones BC, Hawksworth G, Horne VA, et al: Putative active site template model for cytochrome P4502C9 (tolbutamide hydroxylase). Drug Metab Dispos 24(2): 260–266, 1996

Mamiya K, Ieiri I, Shimamoto J, et al: The effects of genetic polymorphisms of CYP2C9 and CYP2C19 on phenytoin metabolism in Japanese adult patients with epilepsy: studies in stereoselective hydroxylation and population pharmacokinetics. Epilepsia 39(12):1317–1323, 1998

Ono S, Hatanaka T, Miyazawa S, et al: Human liver microsomal diazepam metabolism using cDNA-expressed cytochrome P450s: role of CYP2B6, 2C19 and the 3A subfamily. Xenobiotica 26(11):1155–1166, 1996

Potschka H, Fedrowitz M, Loscher W: P-glycoprotein and multidrug resistance-associated protein are involved in the regulation of extracellular levels of the major antiepileptic drug carbamazepine in the brain. Neuroreport 12(16): 3557–3560, 2001

Spina E, Pisani F, Perucca E: Clinically significant pharmacokinetic drug interactions with carbamazepine: an update. Clin Pharmacokinet 31(3):198–214, 1996

Wang JS, Backman JT, Kivisto KT, et al: Effects of metronidazole on midazolam metabolism in vitro and in vivo. Eur J Clin Pharmacol 56(8):555–559, 2000

Weiss ST, Silverman EK, Palmer LJ: Case-control association studies in pharmacogenetics. Pharmacogenomics 1(3):157–158, 2001

Yokogawa K, Shimada T, Higashi Y, et al: Modulation of *mdr1a* and *CYP3A* gene expression in the intestine and liver as possible cause of changes in the cyclosporin A disposition kinetics by dexamethasone. Biochem Pharmacol 63(4): 777–783, 2002

Chapter 8

Test Case

A 41-year-old woman carried the diagnoses of 1) major depressive disorder without psychotic features (both "mood lability" and "obsessional features" noted), 2) anxiety disorder NOS (not otherwise specified), and 3) migraine headaches. She was not responding to her current medication regimen, and a psychopharmacology consultation was requested. Her medications at the time of the consultation were

1. Carbamazepine (Tegretol), 600 mg qhs (blood level=7.5 μg/mL)
2. Topiramate (Topamax), 150 mg/day
3. Clomipramine (Anafranil), 50 mg/day
4. Paroxetine (Paxil), 60 mg/day
5. Clonazepam (Klonopin), 0.5 mg tid
6. Olanzapine (Zyprexa), 10 mg qhs
7. Oral contraceptive medication (Trivora)
8. Percocet (oxycodone+acetaminophen), 2 tabs every 4 hours prn for migraine pain
9. Meperidine (Demerol), 50 mg im every 4 hours prn for migraine pain

The test is to list all the drug interactions contained in this regimen. The answers are given on the next page. After the answers, I've included the recommendations that our team made to the referring psychiatrist.

Answers

The likely interactions contained in this regimen are as follows (summarized in table on opposite page):

1. Carbamazepine inducing the oral contraceptive via 3A4 (Guengerich 1990; Spina et al. 1996)
2. Carbamazepine inducing the clomipramine via 3A4 and 1A2 (Nielsen et al. 1996; Parker et al. 1998; Spina et al. 1996)
3. Carbamazepine inducing the olanzapine via 1A2 and glucuronidation enzyme 1A4 (Hachad et al. 2002; Linnet 2002; Parker et al. 1998)
4. Carbamazepine inducing the meperidine via 3A4 (Piscitelli et al. 2000; Spina et al. 1996)
5. Carbamazepine inducing the topiramate via glucuronidation enzyme 1A4 (Spina et al. 1996)
6. Topiramate inducing the carbamazepine via 3A4 (Benedetti 2000; Spina et al. 1996)
7. Topiramate inducing the oral contraceptive via 3A4 (Rosenfeld et al. 1997)
8. Topiramate inducing the clomipramine via 3A4 (Benedetti 2000; Nielsen et al. 1996)
9. Topiramate inducing the meperidine via 3A4 (Benedetti 2000; Piscitelli et al. 2000)
10. Topiramate inhibiting the metabolism of clomipramine at 2C19 (Anderson 1998; Nielsen et al. 1996)
11. Paroxetine inhibiting the metabolism of clomipramine at 2D6 (Nielsen et al. 1996; von Moltke et al. 1995)

Our team's recommendations were as follows:

1. Discontinue the carbamazepine and topiramate.
2. Begin divalproex sodium (Depakote).
3. Discontinue the paroxetine.
4. Titrate the clomipramine to therapeutic effect and, if needed, consider synergizing with high-dose mirtazapine (Remeron).
5. Liberalize the use of clonazepam.
6. Consider discontinuing the olanzapine.
7. Consider discontinuing the Percocet and meperidine for migraine pain in favor of metoclopramide (Reglan) and/or ketorolac (Toradol). (These drugs had proven effective for this patient in the past.)

		Interaction	Pathway	References
Induction	Carbamazepine	Oral contraceptive	3A4	Guengerich 1990; Spina et al. 1996
		Clomipramine	3A4, 1A2	Nielsen et al. 1996; Parker et al. 1998; Spina et al. 1996
		Olanzapine	1A2, UGT1A4	Hachad et al. 2002; Linnet 2002; Parker et al. 1998
		Meperidine	3A4	Piscitelli et al. 2000; Spina et al. 1996
		Topiramate	UGT1A4	Spina et al. 1996
	Topiramate	Carbamazepine	3A4	Benedetti 2000; Spina et al. 1996
		Oral contraceptive	3A4	Rosenfeld et al. 1997
		Clomipramine	3A4	Benedetti 2000; Nielsen et al. 1996
		Meperidine	3A4	Benedetti 2000; Piscitelli et al. 2000
Inhibition	Topiramate	Metabolism of clomipramine	2C19	Anderson 1998; Nielsen et al. 1996
	Paroxetine	Metabolism of clomipramine	2D6	Nielsen et al. 1996; von Moltke et al. 1995

Note. UGT=uridine 5′-diphosphate glucuronosyltransferase.

References

Anderson GD: A mechanistic approach to antiepileptic drug interactions. Ann Pharmacother 32(5):554–563, 1998

Benedetti MS: Enzyme induction and inhibition by new antiepileptic drugs: a review of human studies. Fundam Clin Pharmacol 14(4):301–319, 2000

Guengerich FP: Metabolism of 17α-ethynylestradiol in humans. Life Sci 47(22): 1981–1988, 1990

Hachad H, Ragueneau-Majlessi I, Levy RH: New antiepileptic drugs: review on drug interactions. Ther Drug Monit 24(1):91–103, 2002

Linnet K: Glucuronidation of olanzapine by cDNA-expressed human UDP-glucuronosyltransferases and human liver microsomes. Hum Psychopharmacol 17(5):233–238, 2002

Nielsen KK, Flinois JP, Beaune P, et al: The biotransformation of clomipramine in vitro: identification of the cytochrome P450s responsible for the separate metabolic pathways. J Pharmacol Exp Ther 277(3):1659–1664, 1996

Parker AC, Pritchard P, Preston T, et al: Induction of CYP1A2 activity by carbamazepine in children using the caffeine breath test. Br J Clin Pharmacol 45(2): 176–178, 1998

Piscitelli SC, Kress DR, Bertz RJ, et al: The effect of ritonavir on the pharmacokinetics of meperidine and normeperidine. Pharmacotherapy 20(5):549–553, 2000

Rosenfeld WE, Doose DR, Walker SA, et al: Effect of topiramate on the pharmacokinetics of an oral contraceptive containing norethindrone and ethinyl estradiol in patients with epilepsy. Epilepsia 38(3):317–323, 1997

Spina E, Pisani F, Perucca E: Clinically significant pharmacokinetic drug interactions with carbamazepine: an update. Clin Pharmacokinet 31(3):198–214, 1996

von Moltke LL, Greenblatt DJ, Court MH, et al: Inhibition of alprazolam and desipramine hydroxylation in vitro by paroxetine and fluvoxamine: comparison with other selective serotonin reuptake inhibitor antidepressants. J Clin Psychopharmacol 15(2):125–131, 1995

P450 Tables

Kelly L. Cozza, M.D.
Scott C. Armstrong, M.D.
Jessica R. Oesterheld, M.D.

Drugs metabolized by 2D6

Antidepressants	**Antipsychotics**	**Other drugs**		
Tricyclic antidepressants[1]	Chlorpromazine	*Analgesics*	*Cardiovascular drugs*[3]	
Amitriptyline	Clozapine[4]	Codeine[5]	Alprenolol	Mexiletine
Clomipramine	Fluphenazine[3]	Hydrocodone	Bufuralol	Nifedipine
Desipramine	Haloperidol[3]	Lidocaine[2]	Carvedilol	Nisoldipine
Doxepin	Perphenazine[3]	Methadone[2]	Diltiazem	Propafenone
Imipramine	Quetiapine[3]	Oxycodone	Encainide	Propranolol[7]
Nortriptyline[2]	Risperidone[3]	Tramadol[6]	Flecainide	Timolol
Trimipramine	Thioridazine[3]		Metoprolol	
Other antidepressants	**Other psychotropics**			
Fluoxetine[3]	Aripiprazole		*Miscellaneous drugs*	
Fluvoxamine[3]	Atomoxetine	Amphetamine	Dextromethorphan[8]	Minaprine
Maprotiline[3]		Benztropine[2]	Donepezil[2]	Ondansetron[2]
Mirtazapine[3]		Cevimeline	Indoramin	Phenformin
Nefazodone		Chlorpheniramine	Loratadine[3]	Tacrine[2]
Paroxetine		Delavirdine[2]	Metoclopramide	Tamoxifen[2]
Sertraline		Dexfenfluramine		
Trazodone[3]				
Venlafaxine[3]				

[1]Tricyclic antidepressants (TCAs) use several enzymes for metabolism. The secondary tricyclics are preferentially metabolized by 2D6, the tertiary tricyclics by 3A4. TCAs are also oxidatively metabolized by 1A2 and 2C19. [2]Oxidatively metabolized primarily by 2D6. [3]Metabolized by other P450 enzymes. [4]2D6 is a minor pathway; 3A4 and 1A2 are more prominent. [5]O-Demethylated to morphine by 2D6, a minor pathway. [6]Metabolized to a more active pain-relieving compound, M1. [7]β-Blockers are partly or primarily metabolized by 2D6. [8]Used as a probe for 2D6 activity.

Inhibitors of 2D6

Antidepressants	***Antipsychotics***	***Other inhibitors***	
Amitriptyline	Chlorpromazine	Amiodarone	Methylphenidate
Bupropion	Clozapine	Chlorpheniramine	**Metoclopramide**
Desipramine	Fluphenazine	Celecoxib	Mibefradil
Fluoxetine	Haloperidol	**Cimetidine**	Pimozide[1]
Fluvoxamine[1]	Perphenazine	Clomipramine	**Quinidine**
Imipramine	Risperidone	Diphenhydramine	**Ritonavir**
Norfluoxetine	Thioridazine	Doxorubicin	Terbinafine
Paroxetine		Lansoprazole	Ticlopidine[1]
Sertraline[2]		Lopinavir/Ritonavir	Valproic acid
Venlafaxine		Loratadine	Yohimbine
		Methadone	

Note. Names of potent inhibitors are in **bold** type.
[1]In vitro evidence exists only for the *potential* for potent inhibition of 2D6.
[2]Sertraline's inhibition seems to be dose specific, with higher doses resulting in more potent inhibition than lower doses.

Drugs metabolized by 3A4		
Antidepressants	***Antipsychotics* (continued)**	***Psychotropic drugs, other***
Amitriptyline[1]	Pimozide[6]	Buspirone
Citalopram[2]	Quetiapine[2]	Donepezil[2]
Clomipramine[1]	Risperidone[7]	Galantamine[2]
Doxepin[1]	Ziprasidone[2,6]	***Other drugs***
Fluoxetine[2]	***Sedative-hypnotics***	*Analgesics*
Imipramine[1]	*Benzodiazepines*	Alfentanil
Mirtazapine[2]	Clonazepam	Buprenorphine
Nefazodone	Diazepam[2]	Codeine (10%, N-demethylated)
Paroxetine[2]	Flunitrazepam[2]	Fentanyl
Reboxetine	Nitrazepam[2]	Hydrocodone[2]
Sertraline[2]	*Triazolobenzodiazepines*	Meperidine[8]
Trazodone[2,3]	Alprazolam	Methadone
Trimipramine[1]	Estazolam	Propoxyphene[9]
Venlafaxine[2]	Midazolam	Sufentanil
Antipsychotics	Triazolam	Tramadol[2]
Aripiprazole	*Other sedative-hypnotics*	*Antiarrhythmics*[6]
Chlorpromazine[2]	Zaleplon	Amiodarone
Clozapine[4]	Zolpidem	Lidocaine
Haloperidol[5]	Zopiclone[2]	Mexiletine[2]
Perphenazine[2]		Propafenone[2]
		Quinidine

Drugs metabolized by 3A4 *(continued)*

Other drugs **(continued)**	***Other drugs*** **(continued)**	***Other drugs*** **(continued)**
Antibiotics (miscellaneous)	*Antimalarials*	*Antiparkinsonian drugs*
Ciprofloxacin	Chloroquine	Bromocriptine
Rifabutin	Halofantrine[6]	Pergolide[2]
Rifampin	Primaquine	Ropinirole[2]
Sparfloxacin[2,6,10]	*Antineoplastics*	Selegiline[2]
Antiepileptics	Bulsulfan	Tolcapone[2]
Carbamazepine	Cyclophosphamide[2]	*Antiprogesterone agents*
Ethosuximide[2]	Daunorubicin	Lilopristone
Felbamate[2]	Docetaxel	Mifepristone
Methsuximide[2]	Doxorubicin	Onapristone
Tiagabine[2]	Etoposide	Toremifene[2]
Valproic acid[2]	Ifosfamide[2]	*Antirejection drugs*
Zonisamide[2]	Paclitaxel[2]	Cyclosporine
Antihistamines	Tamoxifen[2]	Sirolimus (Rapamune)
Astemizole[6,10]	Teniposide	Tacrolimus
Chlorpheniramine	Trofosfamide	β-*Blockers*
Ebastine[6]	Vinblastine	Metoprolol[2]
Loratadine[6,7]	Vincristine	Propranolol[2]
Terfenadine[6,10]	Vindesine	Timolol[2]
	Vinorelbine	

Drugs metabolized by 3A4 *(continued)*

Other drugs **(continued)**	***Other drugs*** **(continued)**	***Other drugs*** **(continued)**
Calcium-channel blockers	*Macrolide/ketolide antibiotics* (continued)	*Steroids*
Amlodipine	Rokitamycin	Cortisol
Diltiazem[2]	Telithromycin	Dexamethasone
Felodipine	Troleandomycin	Estradiol
Nicardipine	*Nonnucleoside reverse transcriptase inhibitors*	Gestodene
Nifedipine	Delavirdine[2]	Hydrocortisone
Nimodipine[2]	Efavirenz	Methylprednisolone
Nitrendipine	Nevirapine[2]	Prednisone
Verapamil[2]	*Protease inhibitors (antivirals)*	Progesterone
HMG-CoA reductase inhibitors[11]	Amprenavir	Testosterone
Atorvastatin	Indinavir	*Triptans*
Cerivastatin[10]	Lopinavir	Almotriptan[2]
Lovastatin	Nelfinavir	Eletriptan
Pravastatin	Ritonavir	
Simvastatin	Saquinavir	
Macrolide/ketolide antibiotics	*Proton pump inhibitors*	
Azithromycin	Esomeprazole[11]	
Clarithromycin	Lansoprazole[11]	
Dirithromycin	Omeprazole[11]	
Erythromycin	Pantoprazole[11]	
	Rabeprazole	

Drugs metabolized by 3A4 *(continued)*

***Other drugs* (continued)**	***Other drugs* (continued)**	***Other drugs* (continued)**
Miscellaneous drugs	*Miscellaneous drugs* (continued)	*Miscellaneous drugs* (continued)
Acetaminophen[2]	Diclofenac[2]	Metoprolol[2]
Carvedilol[2]	Ergots	Miconazole
Cevimeline[2]	Fluconazole	Montelukast
Cilostazol[2]	Itraconazole	Ondansetron[2]
Cisapride[6,10]	Ketoconazole	Sildenafil
Colchicine	Levomethadyl[6]	Sibutramine
Cyclobenzaprine[12]	Meloxicam[2]	Vesnarinone
Dextromethorphan[13]		

Note. HMG-CoA = hydroxymethylglutaryl–coenzyme A.
[1]Tertiary tricyclics are metabolized preferentially by 3A4 but are also metabolized by 1A2, 2C19, 2D6, and uridine 5′-diphosphate glucuronosyltransferases.
[2]Also significantly metabolized by other P450 and/or phase II enzymes.
[3]Metabolized by 3A4 to *m*-chlorophenylpiperazine.
[4]Also metabolized by 1A2 and, to a lesser extent, 2D6.
[5]Also metabolized by 2D6 and 1A2.
[6]Potentially toxic to the cardiac conduction system at high levels and therefore should not be used with potent inhibitors of 3A4.
[7]Also metabolized by 2D6.
[8]Not confirmed but strongly suspected to be a primary metabolic pathway.
[9]3A4 activates to analgesic norpropoxyphene.
[10]No longer available in the United States. [11]Also metabolized by 2C19. [12]Also metabolized by 1A2.
[13]N-Demethylation specific for 3A4; reaction can be a probe for 3A4 activity.

Inhibitors of 3A4

Antidepressants	***Antimicrobials* (continued)**	***Other inhibitors***	
Selective serotonin reuptake inhibitors[1]	*Macrolide and ketolide antibiotics*	Anastrozole	Mifepristone
Fluoxetine	**Clarithromycin**	Androstenedione	Nifedipine
Fluvoxamine	**Erythromycin**	Bromocriptine	Omeprazole
Norfluoxetine	**Telithromycin**	Chloroquine	Oral contraceptives
Paroxetine	**Troleandomycin**	Cimetidine[8]	Phenobarbital
Sertraline	*Nonnucleoside reverse transcriptase inhibitors*	Cisapride	Primaquine
Other antidepressants	**Delavirdine**	Cyclosporine	Propoxyphene
Nefazodone	**Efavirenz**	**Diltiazem**	Tacrolimus
Antimicrobials	*Protease inhibitors*	**Grapefruit juice**[3]	Tamoxifen
Antibiotics, other	Amprenavir	Methadone	Valproic acid
Ciprofloxacin[2]	**Indinavir**	Methylprednisone	Verapamil
Norfloxacin[3]	Lopinavir/Ritonavir[6]	**Mibefradil**[9]	Zafirlukast[10]
Quinupristin/Dalfopristin	Nelfinavir		
Sparfloxacin[3]	**Ritonavir**[7]		
Azole antifungals	Saquinavir		
Fluconazole[4]	***Antipsychotics***		
Itraconazole	Haloperidol		
Ketoconazole[5]	Pimozide		
Miconazole			

Note. Names of potent inhibitors are in **bold** type. [1]Also inhibit other P450 enzymes. [2]Also a potent inhibitor of 1A2. [3]Also an inhibitor of 1A2. [4]Potent inhibitor of 2C9. [5]Also an inhibitor of 2C19. [6]Trade name Kaletra. [7]Also a potent inhibitor of 2D6, 2C9, and 2C19. [8]Also an inhibitor of 2D6, 1A2, and 2C9. [9]Also a potent inhibitor of 2D6 and 1A2; no longer available in the United States. [10]Also an inhibitor of 1A2 and 2C9.

Inducers of 3A4

Antiepileptics	***Other inducers***
Carbamazepine[1]	Cisplatin
Felbamate	Cyclophosphamide
Oxcarbazepine	Dexamethasone
Phenobarbital[1]	Efavirenz
Phenytoin[1]	Ifosfamide
Primidone	Lopinavir/Ritonavir[2]
	Methadone
	Methylprednisolone
	Modafinil
	Nevirapine
	Pioglitazone
	Prednisone
	Rifabutin
	Rifampin[1]
	Rifapentine[1]
	Ritonavir[3]
	St. John's wort
	Troglitazone[4]

Note. Names of potent inhibitors are in **bold** type.
[1]"Pan-inducers"—also induce most other P450 enzymes.
[2]Trade name Kaletra.
[3]Currently known to potently induce only 3A4. [4]Removed from the United States market.

Drugs metabolized by 1A2

Antidepressants	Antipsychotics	Other drugs	
Amitriptyline[1,2]	Chlorpromazine[3]	Acetaminophen	Phenacetin
Clomipramine[1]	**Clozapine**[5]	**Caffeine**	Propafenone[9]
Fluvoxamine[3]	Fluphenazine	**Cyclobenzaprine**	Propranolol[9]
Imipramine[1]	Haloperidol[3]	Dacarbazine	**Riluzole**
Mirtazapine[4]	Mesoridazine[3]	**Flutamide**	**Ropinirole**
	Olanzapine[6]	**Frovatriptan**	**Ropivacaine**
	Perphenazine	Grepafloxacin[8]	**Tacrine**
	Thioridazine[3]	**Melatonin**	**Theophylline**
	Thiothixene[3]	Mexiletine	Toremifene
	Trifluoperazine[3]	Mibefradil[8]	Verapamil[9]
	Ziprasidone[7]	Naproxen	*R*-Warfarin[10]
		Ondansetron	**Zolmitriptan**
			Zolpidem[9]

Note. Names of drugs are in **bold** type if there is evidence that in normal human use of the drugs, at least 50% of enzymatic metabolism is through 1A2.
[1]Tertiary tricyclic antidepressants (TCAs) are demethylated by 1A2 and 3A4. 2D6, 2C9, and 2C19 also metabolize tertiary TCAs.
[2]N-Demethylation may be preferentially done by 2C19.
[3]Metabolized by other P450 enzymes as well.
[4]Also metabolized by 3A4.
[5]Demethylated by 1A2 to norclozapine. Clozapine is metabolized to clozapine-*N*-oxide by 3A4 and, to a lesser extent, 2D6 and others.
[6]Metabolized 30%–40% by 1A2 and some by 2D6, glucuronidated by the glucuronosyltransferase (UGT) 1A4.
[7]1A2 is a minor route of metabolism.
[8]Removed from the United States market.
[9]Contribution of 1A2 metabolism is small.
[10]Weaker pharmacological isomer of racemic warfarin.

Inhibitors of 1A2

Fluoroquinolone antibiotics	***SSRIs***	***Other drugs***	
Ciprofloxacin	**Fluvoxamine**	Anastrozole	Phenacetin
Enoxacin		Caffeine	**Propafenone**
Grepafloxacin		Cimetidine	Ranitidine[2]
Lomefloxacin		Fluphenazine	Rifampin
Norfloxacin		**Flutamide**[1]	Ropinirole[3]
Ofloxacin		Grapefruit juice	Tacrine
Sparfloxacin		Lidocaine	Ticlopidine
		Mexiletine	Tocainide
		Mibefradil	Verapamil
		Nelfinavir	Zafirlukast
		Oral contraceptives	
		Perphenazine	

Note. Names of drugs are in bold type if there is evidence that in normal human use, the drugs are **potent** inhibitors. SSRI = selective serotonin reuptake inhibitor.

[1]Flutamide's primary metabolite is a potent inhibitor of 1A2.
[2]Scant evidence of 1A2 inhibition.
[3]Weak inhibitor.

Inducers of 1A2

Drugs	***Foods***	***Other inducers***
Caffeine	Broccoli	Chronic smoking[2]
Carbamazepine	Brussels sprouts	
Esomeprazole	Cabbage	
Griseofulvin	Cauliflower	
Lansoprazole	Charbroiled foods[1]	
Moricizine		
Omeprazole		
Rifampin		
Ritonavir		

[1]Possibly induce through stimulation of polycyclic aromatic hydrocarbons (PAHs).
[2]Induces through stimulation of PAHs.

Drugs metabolized by 2C9

Angiotensin II blockers	***Hypoglycemics, oral***[2]	***NSAIDs***[3]	***Other drugs***
Irbesartan	*Sulfonylureas*	Celecoxib	Carmustine
Losartan	Glimepiride	Diclofenac	Dapsone
Valsartan	Glipizide	Flurbiprofen	Fluvastatin[4]
	Glyburide	Ibuprofen	Mestranol[5]
Antidepressants	Tolbutamide	Indomethacin	Paclitaxel[1]
Fluoxetine[1]		Ketoprofen	Phenytoin
Sertraline[1]		Mefenamic acid	Tamoxifen
		Meloxicam	Tetrahydrocannabinol
		Naproxen	Torsemide
		Piroxicam	*S*-Warfarin[6]
		Valdecoxib	Zafirlukast
			?Zolpidem[7]

Note. NSAID=nonsteroidal anti-inflammatory drug.
[1]Metabolized by other P450 enzymes as well.
[2]"Glitazones" are metabolized by 2C8 and 3A4.
[3]Also metabolized extensively by phase II enzymes.
[4]The exception to the "statins," most of which are oxidatively metabolized in part or in full by 3A4.
[5]Metabolized by 2C9 to active 17-hydroxyethinylestradiol.
[6]*S*-Warfarin is the more active isomer of warfarin. *R*-warfarin is metabolized by 1A2.
[7]Metabolized mainly by 3A4 and 1A2.

Inhibitors of 2C9		
Selective serotonin reuptake inhibitors	***Other inhibitors***	
Fluoxetine	Amiodarone[1]	Modafinil
Fluvoxamine	Anastrozole	Phenylbutazone
Paroxetine	Cimetidine	Ranitidine
Sertraline	Clopidogrel	**Ritonavir**
	Delavirdine	Sulfamethoxazole
	Efavirenz	**Sulfaphenazole**
	Fluconazole	Sulfinpyrazone
	Fluvastatin	Valproic acid
	Isoniazid	Zafirlukast

Note. Names of potent inhibitors are in **bold** type.
[1]Amiodarone, an inhibitor of 1A2 and 3A4, is an insignificant 2C9 inhibitor. However, its metabolite, desmethylamiodarone, is a clinically relevant 2C9 inhibitor.

Inducers of 2C9		
Carbamazepine	Phenobarbital	Rifapentine
Cyclophosphamide	Phenytoin	Ritonavir
Ethanol	Rifabutin	Secobarbital
Ifosfamide	Rifampin	?Valproic acid

Drugs metabolized by 2C19

Antidepressants	***Barbiturates***	***Proton pump inhibitors***	***Other drugs***
Amitriptyline[1]	Hexobarbital	Esomeprazole[5]	Alprazolam[6]
Citalopram[2]	Mephobarbital	Lansoprazole[5]	Cilostazol
Clomipramine[3]		Omeprazole[5]	Cyclophosphamide[7]
Fluoxetine[3]		Pantoprazole[5]	Diazepam[8]
Imipramine[3]		Rabeprazole[5]	Flunitrazepam[5]
Moclobemide			Ifosfamide[7]
Sertraline[3]			Indomethacin[9]
Trimipramine[3]			Mephenytoin
Venlafaxine[4]			Nelfinavir[5]
			Phenytoin[9]
			Proguanil[10]
			Propranolol
			Teniposide
			Tolbutamide[9]

[1]Amitriptyline is also metabolized by 1A2, 3A4, and 2C19. Poor metabolizers at 2C19 have been shown to have higher amitriptyline levels.
[2]Also metabolized by 2D6 and 3A4.
[3]Also metabolized by other P450 enzymes.
[4]2C19 is a minor enzyme in venlafaxine's metabolism. 2D6 and 3A4 are the major enzymes.
[5]Also metabolized by 3A4.
[6]2C19 is a minor enzyme in alprazolam's metabolism. 3A4 is the major enzyme.
[7]Also metabolized by 2B6 and 3A4.
[8]Diazepam is demethylated by 2C19, but other P450 enzymes and conjugation enzymes are also involved in clearance. Diazepam is also metabolized by 3A4.
[9]2C9 is the major route of metabolism.
[10]Metabolized by 2C19 to the active compound cycloguanil.

Inhibitors of 2C19		
Selective serotonin reuptake inhibitors	***Other drugs***	
Fluoxetine	Cimetidine	Oral contraceptives
Fluvoxamine	Delavirdine	Oxcarbazepine
Norfluoxetine	Efavirenz	Ranitidine
Paroxetine	**Esomeprazole**	**Ritonavir**
	Felbamate	Sulfaphenazole
Other antidepressants	Fluconazole	**Ticlopidine**
Amitriptyline	Indomethacin	Topiramate
Imipramine	Lansoprazole	Tranylcypromine
	Modafinil	Valdecoxib
	Omeprazole	

Note. Names of drugs are in **bold** type if there is evidence of potent inhibition.

Inducers of 2C19		
Carbamazepine	Phenytoin	Rifampin
?Norethindrone	Prednisone	Ritonavir
Phenobarbital	Rifabutin	?Valproic acid

Drugs metabolized by 2E1

Anesthetics[1]	***Other drugs and chemicals***
Enflurane	Acetaminophen[2]
Halothane	Aniline
Isoflurane	Benzene
Methoxyflurane	Capsaicin
Sevoflurane	Carbon tetrachloride[3]
	Chlorzoxazone[4]
	Dacarbazine[3]
	Ethanol[5]
	Ethylene glycol
	Ketones
	Nitrosamines
	Verapamil[6]

[1]2E1 defluorinates these anesthetics.
[2]2E1 is a minor substrate in normal circumstances. In cases of overdose or induction of 2E1, the hepatotoxic metabolite *N*-acetyl-*p*-benzoquinone imine (NAPQI) is created.
[3]Metabolism by 2E1 leads to production of a hepatotoxic metabolite.
[4]Used as a probe for 2E1 activity.
[5]Metabolized by other hepatic and extrahepatic enzymes.
[6]2E1 is a minor enzyme in verapamil's metabolism. 3A4 and 2C8 are more important.

Inhibitors of 2E1
Diethylcarbamate
Disulfiram
Isoniazid
Watercress[1]

Note. Names of drugs are in **bold** type if there is evidence of clinically potent inhibition.
[1]Inhibition possibly due to phenyl isothiocyanate.

Inducers of 2E1	
Ethanol[1]	Smoking
Isoniazid	Starvation
Obesity	Uncontrolled diabetes
Retinoids	

[1]Chronic ethanol use induces 2E1.

2B6 substrates, inhibitors, and inducers

Substrates	Inhibitors	Inducers
Bupropion	Efavirenz	Cyclophosphamide
Cyclophosphamide[1]	Fluoxetine	Phenobarbital
Diazepam[2]	Fluvoxamine	
Ifosfamide[1]	Nelfinavir	
Nicotine[3]	Orphenadrine	
Propofol	**Paroxetine**	
Sertraline	Ritonavir	
Tamoxifen[4]	Thiotepa	

[1] 2B6 metabolizes this alkylating agent to its active drug.
[2] 2B6 is a minor enzyme; other P450 enzymes and phase II are more important.
[3] 2B6 is secondary to 2A6 in C-oxidation.
[4] Metabolized by 2B6, 2C9, and 2D6 to a potent active antiestrogenic compound.

Appendix B

Uridine 5′-Diphosphate Glucuronosyltransferase (UGT) or Phase II (Glucuronidation) Tables

Jessica R. Oesterheld, M.D.

The content of these tables was obtained from the P450+ Web site http://www.mhc.com/ Cytochromes, run by Jessica R. Oesterheld, M.D., and David N. Osser, M.D. Used with permission from Dr. Jessica Oesterheld.

UGT1A substrates, inhibitors, and inducers

	1A1	1A3	1A4	1A6	1A9
Chromosome	2	2	2	2	2
Polymorphism	CN-I, CN-II, GS			Yes	
Some endogenous substrates	Bilirubin Estriol	Estrones	Androsterone Progestins	Serotonin	2-Hydroxyestradiols Thyroxine
Some substrate drugs	Acetaminophen Atorvastatin Buprenorphine Cerivastatin Ciprofibrate Clofibrate Ethinyl estradiol Flutamide metabolite Gemfibrozil *Nalorphine* *Naltrexone* Simvastatin SN-38 *Telmisartan* *Troglitazone*	*Amitriptyline* Atorvastatin Buprenorphine Cerivastatin *Chlorpromazine* *Clozapine* Cyproheptadine Diclofenac Diflunisal *Diphenhydramine* *Doxepin* Fenoprofen Gemfibrozil *4-Hydroxytamoxifen* Ibuprofen *Imipramine*	*Amitriptyline* Chlorpromazine Clozapine/Desmethyl metabolites Cyproheptadine Diphenhydramine Doxepin 4-Hydroxytamoxifen Imipramine Lamotrigine Loxapine Meperidine Olanzapine Promethazine Retigabine	Acetaminophen Entacapone Flutamide metabolite Ketoprofen Naftazone *SN-38*	Acetaminophen Clofibric acid Dapsone Diclofenac Diflunisal *Ethinyl estradiol* Estrone Flavonoids *Furosemide* *Ibuprofen* *Ketoprofen* Labetalol *Mefenamic acid* *Naproxen* *R*-Oxazepam Propofol

UGT1A substrates, inhibitors, and inducers *(continued)*

	1A1	1A3	1A4	1A6	1A9
		Losartan			Propranolol
		Loxapine			Retinoic acid
		Morphine			*SN-38*
		Nalorphine			Tolcapone
		Naloxone			*Valproate*
		Naltrexone			
		Naproxen			
		Naringenin			
		Norbuprenorphine			
		Promethazine			
		Simvastatin			
		SN-38			
		Tripelennamine			
		Valproate			
Some inhibitors	Tacrolimus		Probenecid	Silymarin (milk thistle)	Cyclosporin A
			Sertraline	Troglitazone	Diflusinal
			Valproate		Flufenamic acid
					Mefanamic acid
					Neflumic acid
					Silymarin (milk thistle)
					Tacrolimus

UGT1A substrates, inhibitors, and inducers *(continued)*

	1A1	1A3	1A4	1A6	1A9
Some inducers	Clofibrate Dexamethasone Flavonoids Phenobarbital Phenytoin Ritonavir		Carbamazepine Ethinyl estradiol Methsuximide Oxcarbazepine Phenobarbital Phenytoin Primidone Rifampin	Dexamethasone 3-Methylcholanthrene (3-MC) Polyaromatic hydrocarbons (PAHs)	Polyaromatic hydrocarbons (PAHs)

Note. *Italics* indicate that the UGT is a minor pathway for the substrate.
CN-I=Crigler-Najjar syndrome type I; CN-II=Crigler-Najjar syndrome type II; GS=Gilbert syndrome; UGT=uridine 5′-diphosphate glucuronosyltransferase.

UGT2B substrates, inhibitors, and inducers

	2B7		2B15
Chromosome	4		4
Polymorphism	Yes		Yes
Some endogenous substrates	Androsterone Bile acid		Catechol estrogens 2-Hydroxyestrone Testosterone
Some substrate drugs	Chloramphenicol Clofibric acid Codeine Cyclosporine Diclofenac *Entacapone* Epirubicin *Fenoprofen* Hydromorphone *Ibuprofen* Ketoprofen Lorazepam Losartan Morphine	Nalorphine Naloxone Naltrexone Naproxen Norcodeine *R*-Oxazepam Oxycodone Tacrolimus Temazepam *Tolcapone* Valproate Zidovudine Zomepirac	Dienestrol *Entacapone* *S*-Oxazepam Phenytoin metabolites *Tolcapone*

UGT2B substrates, inhibitors, and inducers *(continued)*

	2B7		**2B15**
Some inhibitors	Amitriptyline	Methadone	
	Chloramphenicol	Morphine	
	Clofibrate	Naproxen	
	Codeine	Oxazepam	
	Diazepam	Probenecid	
	Diclofenac	Propofol	
	Fenoprofen	Ranitidine	
	Fluconazole	Temazepam	
	Guanfacine	Trimethoprim	
	Ibuprofen	Valproate	
	Ketoprofen		
Some inducers	Ganciclovir	Rifampin	
	Phenobarbital	Tobacco smoking	

Note. *Italics* indicate that the UGT is a minor pathway for the substrate.
CN-I=Crigler-Najjar syndrome type I; CN-II=Crigler-Najjar syndrome type II; UGT=uridine 5′-diphosphate glucuronosyltransferase.

P-Glycoprotein Table

Jessica R. Oesterheld, M.D.

The content of these tables was obtained from the P450+ Web site http://www.mhc.com/Cytochromes, run by Jessica R. Oesterheld, M.D., and David N. Osser, M.D. Used with permission from Dr. Jessica Oesterheld.

Some P-glycoprotein nonsubstrates, substrates, inhibitors, and inducers

Nonsubstrates	Substrates	Inhibitors	Inducers
Alfentanil	Aldosterone	Amiodarone	Dexamethasone
Amantadine	Amitriptyline	Amitriptyline	Doxorubicin
Chlorpheniramine	Amoxicillin	Atorvastatin	?Nefazodone (chronic)
Citalopram	Amprenavir	Bromocriptine	Phenobarbital
Clozapine	Carbamazepine	Chloroquine	Prazosin
Fentanyl	Chloroquine	Chlorpromazine	Rifampin
Fluconazole	Cimetidine	Clarithromycin	Ritonavir (chronic)
Flunitrazepam	Ciprofloxacin	Cyclosporine	St. John's wort
Fluoxetine	Colchicine	Cyproheptadine	Trazodone
Haloperidol	Corticosteroids	Desipramine	?Venlafaxine
Itraconazole	Cyclosporine	Diltiazem	
Ketoconazole	Digitoxin	Erythromycin	
Lidocaine	Digoxin	Felodipine	
Methotrexate	Diltiazem	Fentanyl	
Midazolam	Docetaxel	Fluphenazine	
Sumatriptan	L-Dopa	Garlic	
Yohimbine	Doxorubicin	Grapefruit juice	
	Enoxacin	Green tea (catechins)	
	Erythromycin	Haloperidol	
	Estradiol	Hydrocortisone	
	Fexofenadine	Hydroxyzine	
	Grepafloxacin	Imipramine	
	Indinavir	Itraconazole	
	Irinotecan	Ketoconazole	

Some P-glycoprotein nonsubstrates, substrates, inhibitors, and inducers *(continued)*

Nonsubstrates	Substrates	Inhibitors	Inducers
	Lansoprazole	Lansoprazole	
	Loperamide	Lidocaine	
	Losartan	Lovastatin	
	Lovastatin	Maprotiline	
	Mibefradil	Methadone	
	Morphine	Mibefradil	
	Nelfinavir	Midazolam	
	Nortriptyline	Nefazodone (acute)	
	Ondansetron	Nelfinavir	
	Phenytoin	Ofloxacin	
	Quetiapine	Omeprazole	
	Quinidine	Orange juice (Seville)	
	Ranitidine	Pantoprazole	
	Rifampin	Phenothiazines	
	Ritonavir	Pimozide	
	Saquinavir	Piperine	
	Tacrolimus	Probenecid	
	Talinolol	Progesterone	
	Teniposide	Propafenone	
	Terfenadine	Propranolol	
	Vinblastine	Quinidine	
	Vincristine	Ritonavir (initial)	
		Saquinavir	
		Simvastatin	

Some P-glycoprotein nonsubstrates, substrates, inhibitors, and inducers *(continued)*

Nonsubstrates	Substrates	Inhibitors	Inducers
		Spironolactone	
		Tamoxifen	
		Terfenadine	
		Testosterone	
		Trifluoperazine	
		Valspodar	
		Verapamil	
		Vinblastine	
		Vitamin E	

Subject Index

Page numbers printed in ***boldface*** *type refer to tables.*

Case Index by Drug Interaction Pattern